AF386033

Praise for
Life at the Speed of Play

"The best founders build from first principles. Mark has done that his entire career. He believes great products are shaped by taste, speed, and the courage to act. *Life at the Speed of Play* is essential reading for anyone who wants to build products that people don't just use, but depend on."

—Brian Chesky, CEO of Airbnb

"Mark has taught our YC founders that intellectual honesty is the foundation of a successful company. You have to be willing to kill your losing ideas to make something people want. Read this book and change your odds of success."

—Garry Tan, president of Y Combinator (YC)

"Today, the only bottleneck to building great products is knowing what to create. Mark is an expert at this."

—Sam Altman, CEO of OpenAI

"Mark is a pioneer, one of the absolute greats at building products that people love to use. His ethos and ideas are mandatory reading for anyone trying to craft top-tier consumer experiences."

—Shayne Coplan, CEO of Polymarket

"Mark's insight that your instincts are likely right, but your ideas are likely wrong—and how to manage that conflict—is invaluable. His framework for innovating on entirely new surface areas is important and timely. A must-read for all founders."

—Fred Wilson, venture capitalist and cofounder of Union Square Ventures

"Most product books tell you to move fast and break things. Mark Pincus extends that idea, challenging you to build 'failure machines' and build your ideas at a velocity that wasn't possible until now. This is the playbook for anyone building in the age of AI. Plus it includes a killer Bolognese recipe."

—Ethan Mollick, associate professor at the Wharton School, AI researcher, and *New York Times* bestselling author of the *Co-Intelligence*

"Ambitious founders strive to create Internet Treasures: products so essential we can't imagine life without them. Mark Pincus reveals how his relentless focus on user experience drove Zynga's audacious growth. *Life at the Speed of Play* is Mark's operating manual for building legacy companies, from objectives and key results (OKRs) to Bold Beats and weekly roadmaps."

—John Doerr, venture capitalist and chairman of Kleiner Perkins

LIFE AT THE SPEED OF PLAY

Mark Pincus

HARPER
BUSINESS

An Imprint of HarperCollins*Publishers*

LIFE AT THE SPEED OF PLAY

Launch Products People Love!

hc.com

FIRST EDITION

Designed by Kyle O'Brien

Unless otherwise noted, illustrations and photographs are provided by the author.

Emoji Art © Cali6ro/Shutterstock

Library of Congress Cataloging-in-Publication Data is available upon request.

ISBN 978-0-06-335257-5

Printed in the United States of America

26 27 28 29 30 LBC 5 4 3 2 1

This is a work of nonfiction. The events and experiences detailed herein are all true and have been faithfully rendered as remembered by the author, to the best of his ability, or as they were told to the author by people who were present.

You're living your life in a cage with golden bars. The door is open, but you don't even know it.

—My interpretation of a line in the screenplay
Barfly by Charles Bukowski

Contents

Idea Generators

Reid Hoffman

Mark and I met in 2000, when the Web 1.0 boom was quickly spiraling into an internet winter. I was an executive at PayPal and Mark had just taken Support.com public, but the internet's future seemed anything but certain. As failed start-ups consigned their Aeron chairs to online liquidators (yeah, it was a little ironic), we bonded over our belief that this was just a temporary collapse—not the finished state of the internet at all, but rather just a pause before its evolution into something far more dynamic and essential to human interaction: Web 2.0.

The theory of the game we shared? The future of the web would not be driven by static pages, hierarchical directories, and newspaper articles linked to encyclopedias. Instead, it would revolve around human connection, in ways that would spill over into, then merge with, real life itself. After some false starts and technical constraints, the web would finally evolve into a much more explicitly social medium, oriented around new networks and platforms that prioritized human connection and affiliation.

Along with our shared view that the web was not only not dead, but actually on the verge of massive growth and transformation, Mark most amazed me with the rate at which he generated new ideas. As many Web 1.0 veterans retreated to the seemingly less volatile industries of the past, Mark kept his focus on how to build for the internet's people-centered potential. And he had ideas and theories for seemingly everything: how news media, movies, games, and even politics would evolve in this new coming world; how trust would be gained and lost; which entrenched institutions would sink; which new ones might take root.

As delightful and instructive as it was just to sit and theorize about the world with Mark, though, I also came to realize fairly quickly that his way to intellectually engage with the world was not just to talk but also to act—to test, play, learn, and refine. To that end, he has created and continues to develop a very intentional system for making sense of the world from an entrepreneur's perspective—a set of rules for how to develop, test, and deploy products, brands, and services.

Indeed, as much as Mark is an idea generator, he's also an idea *executor*—and I mean that in two ways. First, as you'll learn in this book, his ideas are always linked to strategies and tactics for making them tangible. Second, as you'll also learn here, he is a strong believer in the premise that most ideas are bad ones—and that ultimately what distinguishes great entrepreneurs and founders is their synergistic capacity to generate and kill ideas with equal speed and enthusiasm. (Check out his podcast episode on *Masters of Scale!*)

Unsurprisingly, given the title of this book and Mark's status as the founder of Zynga, the company that made mainstream social gaming a thing, it's not a metaphor of any kind to say that Mark approaches entrepreneurism like a game.

I love this approach. When I meet founders and product managers, I typically ask them two questions:

What game are you playing?
What's your theory of the game?

The first question—What game are you playing?—is about clarity of vision and strategic positioning. Are you competing in mobile gaming or enterprise SaaS? Are you building for Gen Z on TikTok or professionals on LinkedIn? Too many founders think they're playing one game when they're actually playing another—or, worse, they're trying to play multiple games simultaneously without focusing their resources to win at any one of them. Understanding what game you're playing means understanding your arena, your competitors, your constraints, and, most important, your win conditions. It means knowing whether you're playing for market share, profitability, user engagement, or network effects—and having the discipline to optimize for that specific outcome.

The second question goes deeper. Your theory of the game is your mental model of how success happens in a given domain. What are the dynamics that underlie it? What does everyone else playing in this arena believe that might be wrong—and how can you use that to create a compelling and sustainable advantage? A theory of the game is your hypothesis about cause and effect, timing, and sequencing. It makes the difference between randomly trying things and having a strategic framework that guides your decisions. The best founders don't just have a theory; they can articulate it, test it, and refine it based on what they learn.

These questions matter because games, in their essence, are a particular kind of map—a reductive, constrained version of the world that paradoxically helps us navigate reality more effectively. Within the bounded spaces of games, the often muddled complexities of reality are distilled into something manageable and clear. The rules and goals are explicit. So is the feedback, and it starts almost immediately. In a well-designed game, it doesn't take long for you to understand whether you're winning or losing.

All this clarity also creates efficiency. In real life, you often expend a great deal of energy trying to figure out what the rules are, or litigating whether your opponents are playing fairly, or questioning what you're even trying to accomplish. In a game, you just play. The beauty of games—and the reason so many entrepreneurs are drawn to them—is that the skills you develop translate to real-world situations that are far murkier and less responsive. You learn to recognize patterns, adapt strategies, and make decisions under uncertainty in an environment that gives you immediate feedback.

Finally, when you have an explicit theory of the game, you can build a company around it; it creates a framework for win conditions, metrics for progress toward success, and decomposition of goals and work processes.

The systemic practices Mark describes in this book are effectively the meta-game, a toolbox of approaches you can use when pursuing your own theory of the game. Drawing from his experience as a serial founder who scored big wins across the web, mobile, and now AI, Mark shares key concepts and truths about human motivation, feedback loops, and social dynamics that apply to virtually any context where your aim is to create some product or service that others will not only find useful but want to regularly keep using over time.

Indeed, one way to summarize Mark's career is that in the process of building Zynga and many other start-ups, he's gamified entrepreneurship, product development, and company building itself. This book is the set of rules—or perhaps some cheat codes—for the game of rapidly building products, platforms, and companies, in ways that are designed to help them endure for the long term.

His perspective and frameworks are more relevant now than they've ever been, for two interconnected reasons, both relating to AI. First, platform shifts represent the best times to put frameworks like Mark's into action. When the rules are changing, when new capabilities emerge, when distribution channels evolve—these are the moments when a strong theory of the game matters most. Operating in uncertainty is precisely when a theory serves as your compass. It doesn't tell you exactly where to go, but it orients you. It helps you interpret ambiguous signals, make decisions with incomplete information, and recognize patterns in the chaos. The key is understanding that your theory must be a living thing—something you're fully invested in learning from, adapting, and iterating on. And we're living through perhaps the most significant platform shift in computing history.

Second, this particular platform shift—AI—effectively gives everyone the capabilities to be a little more like Mark (including Mark himself). We can all now generate torrents of ideas in an ongoing fashion. We can test them out faster than ever before through vibe coding, text-to-image generation, video synthesis, and audio creation. As much as the early web democratized media, it still took considerable resources to launch a professionally competitive website in the Web 1.0 era. Now, however, you can prototype websites, apps, games, and even entire platforms in a few hours.

This democratization of creation, this new form for individual and collective superagency, makes Mark's frameworks massively more valuable.

When everyone can generate and build, the competitive advantage shifts; instead of "Can you build it?," the question becomes "Do you know what to build, how to test it, and when to kill it?"

Having a strategy to rapidly test ideas, identify the promising ones, and ruthlessly eliminate the rest isn't just valuable—it's essential. And that's exactly what this book gives you.

Read it. Learn the rules. Develop your theory. Then get in the game.

Introduction

Internet Treasures

My good friend and lifelong collaborator Reid Hoffman calls me an idea generator, but my challenge has always been bringing my ideas to life.

That is true for all of us. There are so many ideas that we never pursue. Real life can often feel like a beatdown. Once we get to imagining the many steps necessary to actually implement our idea, it can feel overwhelming and expensive. It feels less fun, and we decide that maybe the idea wasn't that great. Even when we do pursue our idea, we're forced to make many painful trade-offs, with time as our primary currency.

The way to change this equation is to massively accelerate the speed at which we build our ideas.

What if we could all live our lives at the speed of play?

Life at the speed of play is a mindset that enables us to turn our ideas into reality with the same ease, speed, and freedom that we experience in games or play. We can test our ideas in hours or days rather than months or years.

Play offers us a way to escape reality's constraints. In Monopoly, you can build a business empire and then go totally broke. On Halloween, you can try on any persona for a night. When you play, you can explore wild ideas without worrying about failure.

Elon already lives life at the speed of play. He says there's a high probability we are living in a simulation. Elon seems to have beaten the sim and is having way more fun—from fart mode in his cars to flamethrowers, perfume, and in summer 2025, the launch of Tesla diners. In December 2016, he tweeted, "Traffic is driving me nuts. Am going to build a tunnel boring machine and start digging." Within a month, he raised $1 billion and launched The Boring Company.

Imagine bringing that whimsical, playful, and fast approach to how *we* build products. Think how many more ideas we'd test if we didn't worry about costs and consequences. The fear of bad decisions holds us back from taking risks.

Building products at the speed of play means we can explore 100 or even 1,000 ideas in the time it used to take to pursue one. It's not just about failing fast; it's about expanding our zones of innovation. By giving ourselves permission to test even our weirdest ideas, our range of innovation expands massively, and so does our learning—allowing us the opportunity to get to breakout ideas.

At Zynga, we lived life at the speed of play. We were committed to testing more ideas in a week than the game industry tested in a year. By cutting the marginal cost of a test to nearly zero, we operated in a state where any idea, no matter how wacky, was a potential shot on goal. In YoVille, we tested and quickly launched avatar babies, which lit up our player base. We called these **OMFG moments**.

As entrepreneurs and product makers, we're obsessed with speed, but most of us are doing it wrong. We've been sold on Minimum Viable Product (MVP), but the world moves too fast for that now. What we need is a much smaller **atomic unit** of our idea that enables us to test and fail fast, long before we build the full product.

When we were creating games at Zynga, it wasn't about product market *fit*. It was about product market *hits*. In fact, we managed to generate an 80 percent hit rate on our major games. We launched eight games that achieved more than $1 million in daily revenue and together generated more than one billion app installs and $1 billion in revenue. Meanwhile, I would estimate that a well-executed consumer app has a 1 percent chance of any success, and an exceptional one might have a 10 percent chance. What if we could multiply those odds by a factor of eight?

This book highlights the most important lessons from my 30-year journey as a founder and product maker. It's based on two Stanford courses I've created and taught. But think of it as more of a dinner conversation than a lecture. I'm sharing my pattern recognition from decades of trying, failing, and sometimes succeeding to help you turn your winning instincts into hit products.

At its core, this book centers on three pillars that are crucial for any product maker:

1. **Isolating your winning instincts from your losing ideas.** Learn why your instinct is right 95 percent of the time, but your idea is wrong at least 75 percent of the time—and what to do about that.
2. **Turning insights into reality using my Proven Better New framework.** Turn your ideas into heat-seeking missiles that find their target faster with fewer resources. Make every engineering day count.
3. **Getting people to do the right thing when you're not in the room.** Build a culture and system that can repeatedly deliver breakthrough products.

Finally, we'll talk about how to apply all of these principles in a world that is being reshaped by new technologies. It's clear AI alone will reshape our world, but robotics and space exploration might redefine everything.

This Is God's Work

This generation of product makers has an unprecedented opportunity to build what John Doerr, chair of Kleiner Perkins, and Bing Gordon, my longtime friend and coach and the former Chief Creative Officer of Electronic Arts (EA), call *Internet Treasures*—services we can't remember life before or imagine life without. Doerr and Gordon believe that one day, these products will be shown in the Smithsonian as the greatest output of our culture: the iPhone, Amazon, Google. Now we can add ChatGPT, too.

When I first heard about Internet Treasures, I was finally able to put words around my "why" and what my career has been about. I truly believe this pursuit—turning our instincts into products that can enhance millions of people's lives—is God's work, the most important thing we, as product makers, can contribute. That may sound like lofty talk from a guy who made video games, but we succeeded because they delivered so much more than just entertainment. Bing once said we should measure the success of our social games in marriages—and, in fact, there were many. This past year, two students approached me after my Stanford class to share that they first flirted in Words With Friends. Now they're married.

When we get our product right, it feels like a communion with our users. We know it in our gut, and we feel it in the way our users connect with us. That's why, for me, the true measurement of our success has always been player hugs.

Being a great product maker isn't about building one perfect product. It's about developing a systematic approach that increases your odds of a breakout hit. I think of us as a community of product managers (PMs) chasing the same dragon. Some of us are just earlier in our careers, but we're all on the same path and after the same prize.

At some point, I started jokingly calling this playbook my "PM Bible." Not because it contains absolute truths, but because it's something I return to again and again to keep me and my teams intellectually honest and accountable. When we're caught up in the excitement of building something new or emotionally attached to an idea, these principles bring us back to our true path.

One of my biggest motivations to write this book has been the charge I get every time I run into an ex–Zynga PM or a founder who has implemented my frameworks to tune into their own winning instincts. Shayne Coplan, the founder of Polymarket, taught himself *Proven Better New* from my YouTube videos before we ever met. As of November 2025, Polymarket was valued at $9 billion!

Polymarket is one great example of how these principles apply far beyond games, to any consumer application. Later, we'll talk about how these principles have even played out in enterprise software.

Whether you're a founder, product manager, or anyone involved in creating products, my hope is that this book will help you navigate your own journey more successfully. The five years my coauthor Carlye and I have spent on this book project will all be worth it if one reader builds a world-changing product.

My Origin Story

I always looked up to my dad and was inspired by the way he lived life—all in on everything he did. He pioneered a category called financial public relations and built a nationwide firm called the Financial Relations Board. He built this firm over 38 years and eventually sold it to an advertising con-glomerate for $38 million. More importantly, my dad lived and modeled a life about play. He was a talented jazz musician, wrote speeches for presidents, and authored many books. Above all, he exhibited an amazing commitment to playfulness, which included writing a (bad) joke book under the pseudonym Roderick Rosebuns (pink ass for Pincus). He even insisted that his gravestone inscription be "third-string quarterback" (referencing his brief, failed high school football career). He created a family culture that was rooted in games and creative expression. Our nights, weekends, and holidays were a constant stream of charades, board games, and eventually poker. For each of these games, our family modified the rules to make them more fun and, usually, faster. When we played Pictionary, we threw away the board and just read each other the questions. With charades, we invented the "moo" rule, which penalized people for making noises. My favorite change was in Scrabble, where my dad made up a rule that enabled you to use another player's letters to change any word on the board, freeing up new letters to get around bad luck of the draw on letters. The single word that summed up my dad the best was "chutzpah"—he got the most out of life while also challenging and often breaking rules if they didn't make sense.

Because my dad was such a strong personality and he taught me to question authority, we had a falling-out when I was 18. He decided he wasn't done raising me and was going to pull me out of college for a year to "fin-ish the job." Of course, as a rebellious teen, my response was "fuck you," and I went on to transfer from the University of Michigan to Wharton and become a straight-A student while getting a lot of jobs to support myself.

Ironically, this drove me to be the kind of self-sufficient high achiever my dad was hoping to raise.

This tendency to make up my own rules, combined with a sense of playfulness, proved to be career limiting as I entered the corporate world, where being a perfect trained poodle was valued far greater than being a mini-CEO. (This period is where I learned that the true meaning of the word "corporate" is lack of thought or intent. Imagine if someone told you, "You seem really corporate." Would you feel flattered or insulted?)

This led to a series of jobs in my twenties where I briefly touched greatness, working for some amazing people, even if I couldn't hold the positions for very long. I was hired by Bill Loomis at the investment bank Lazard Frères and worked for Steve Rattner and Ken Jacobs, all of whom were legends in the banking world. I had the distinction of being one of two Bain summer associates who were asked to leave—but did get to present to Mitt Romney. A career highlight was getting to work for John Malone at TCI, which was at the time the largest cable company in the country. My recollection was that I made some career-limiting moves. It's funny that he later recalled that I was a brilliant hire they were sorry to see leave. My last job working for someone else was for future senator Mark Warner at Columbia Capital, his venture capital (VC) firm in DC. In each of these roles, I was exposed to strong leaders, but it didn't feel like there was room for me to grow into one myself. I had a deep desire to be a CEO and started to realize the importance of turning everyone else into one, too.

As I reflect on that period, I realize that my actual career started on my first day as a product manager in 1994 at a company called SpaceWorks in Bethesda, Maryland. I had sourced this investment for Columbia Capital, which, luckily for me, had a strategy of embedding its associates in its portfolio companies. In one sense, SpaceWorks was a dream job, because I got to experience the high of building and shipping real products. However, I quickly soured on the investment as I saw how out of position the company was. SpaceWorks was building private online networks for big enterprises at the exact moment that the public web was starting to take off. Space-Works was painfully out of position, and there was nothing I could do about it. I tried to convince the CEO and the partners at Columbia Capital that SpaceWorks needed a hard pivot, but nobody would listen. I knew the only way I was going to have a chance to prove I was right was to be a CEO, and that meant starting my own company.

While SpaceWorks ultimately failed, the opportunity was life-changing. In addition to launching my career as a product maker, I met two brilliant young engineers, Cadir Lee and Scott Dale, who would become my co-founders at Support.com and later join me in building Zynga.

In the fall of 1995, I left Columbia Capital to cofound my first company, FreeLoader, with Sunil Paul. FreeLoader pioneered push technology with free software that automatically downloaded websites and made them available through an interactive screensaver. Through a combination of intense hustle and luck, we built and sold the company in seven months for $38 million. It was not lost on me or my dad that this was the exact same dollar amount he sold his company for, only after 38 years! This was my first taste of living life at the speed of play.

Despite FreeLoader's successful exit, and all the accolades, it felt hollow for me. FreeLoader, under the acquiring company Individual Inc., died within a year, as our technology became less relevant and we didn't have the autonomy or operating budget to pivot. After this experience, I knew I wanted to build something enduring, like my father had.

In 1996, Sunil and I moved the company—and our lives—to San Francisco, because we both thought it was going to be the Motor City of the internet (like Detroit was for the car industry). By 1997, I was back at it, cofounding Support.com with Cadir and Scott. Despite having no experience in enterprise software, we built Support.com into the leader in helpdesk automation software. We were successful so quickly, and I was still so young at 30, that my VCs felt the need to replace me as CEO with someone gray and corporate before we could go public, which we eventually did, in 2000, at a valuation of $1.5 billion.

In 2001, after the new Support.com CEO pushed me out for undermining her authority and going after a large deal that would require a different product direction, I cofounded a product incubator with my friend Martin Roscheisen called Tank Hill, named after the hill behind my house in Cole Valley where I walked my dog Zinga every day. We quickly saw that the consumer internet was about to be over, maybe for a long time, so we shut it down and returned $7 million of the $10 million we had raised. I like to think that founders should be good stewards in down markets, too. Sometimes failing fast and returning capital is the best, maybe even only, option.

After this, I entered the first of what I like to call my *Abysses*—the open space so many of us founders experience in between building companies. This lasted a few years. More on Abysses later.

In 2003, I cofounded Tribe.net with Paul Martino and Val Syme. Tribe, which was one of the first three social networks, was focused on combining Craigslist (classified listings) with Friendster (social networking). It turned out to be a long, painful failure—we couldn't find product-market fit, a private label licensing deal with Cisco went nowhere, and we eventually ran out of funding—but it generated learning that led to the founding of my next company, Zynga. It also led to my $38K seed investment in Facebook (now Meta).

In 2007, I founded Zynga, which was named after my amazing American bulldog. I saw in Zynga the opportunity to finally build my forever company. This is when I realized my career was about building an Internet Treasure, and I saw in Zynga the chance to do that. Since we were profitable from the beginning, I was able to keep control this time, which meant I could build a house I would want to live in—I could make decisions for the long term and not worry about pleasing VCs. Zynga pioneered social gaming and helped turn play into a mass-market activity. We achieved $1 billion in revenue within four years and went public in 2011. In 2012, as a newly public company, Zynga lost 30 percent of its traffic in a few days, after Facebook changed its algorithms. This led to a painful turnaround as we rebuilt our products and company for mobile. During that time, I left and returned as CEO before eventually partnering with current CEO, Frank Gibeau. In 2022, Zynga merged with Take Two Interactive (best known for its hit franchise *Grand Theft Auto*) in a merger that valued the company at $12.7 billion.

In 2018, after feeling confident that Frank and team could run Zynga without me, I cofounded an investment firm called Reinvent Capital with Reid Hoffman and Michael Thompson. Our focus was doing venture capital at scale. We backed some amazing founders and companies like SpaceX and Joby, which is leading the way to air taxis in a category called electric vertical take-off and landing (eVTOL). We sponsored three initial public offerings (IPOs) for Joby, Aurora (the leader in self-driving trucks), and Hippo (the leading online insurance company).

In 2022, I founded Erth.AI to pursue my vision for the metaverse—delivering on the promise of life at the speed of play. This vision to create an open virtual world that merges games with real life (money and social) had haunted me for the previous 16 years. This project has gone through several iterations as I've explored new technology, first crypto and now AI. Luckily, my team has remained scrappy, which has provided the staying power to keep at this as regulations and technology platforms become clearer.

Me and Zinga the dog—Zynga's namesake!
Credit: Jim Wilson /*The New York Times*/Redux

My Career TL;DR

1988: Graduate Wharton, join Lazard Frères (investment bank)

1993: Graduate Harvard Business School, join Tele-Communications Inc. (TCI) and Liberty Media (cable company)

1994: Join Columbia Capital (venture capital firm)

1995: Cofound FreeLoader with Sunil Paul

1996: Sell FreeLoader for $38 million

1997: Cofound Support.com with Cadir Lee and Scott Dale

2000: Support.com IPO

2000: Cofound Tank Hill (incubator) with Martin Roscheisen

2001: Shut down Tank Hill, return investor money

2001–2003: First Abyss

2003: Cofound Tribe.net with Paul Martino and Val Syme

2006: Shut down Tribe.net

2005–2007: Second Abyss

2007: Found Zynga

2011: Zynga IPO

2014: Found Superlabs (incubator)

2015: Return as Zynga CEO

2018: Cofound Reinvent Capital with Reid Hoffman and Michael Thompson

2022: Take-Two acquires Zynga for $12.7 billion

2022: Found Erth.AI (formerly known as Dot Earth)

My objective in writing this book is to share the lessons and practices I've developed over my 30-year career as a product maker and founder. The book is also a deep dive into the underlying philosophies that have driven me, inspired me, and have made me successful.

At the root of these philosophies is a commitment to intellectual honesty. For me, that all began when I started my Book of Life.

1

Book of Life

Know your goal or suffer death by 1,000 compromises.

At 28, I found myself washed up and living in a junior one-bedroom apartment in Washington, DC. I had been the only kid in my Harvard Business School section to graduate without a job. Over the years, my classmates had become partners at Goldman and McKinsey or had started hedge funds, while I managed to get fired from (or was asked to leave) every place I worked. I was now being pushed out of Columbia Capital, a thenfledgling venture capital firm run by future-senator Mark Warner. I spent my afternoons playing pickup soccer and watching HBO, leaving the office by 4 p.m. At night, I drank beers at the Zoo Bar with my best friend Tom Cole—who was living on my sofa—and a bunch of government mainframe engineers, all of us complaining about our lame jobs.

I found myself in a synagogue for the first time since my bar mitzvah. During Rosh Hashanah, the Jewish New Year, I sat listening to Hebrew prayers I couldn't follow and started writing in a notebook. Over the next 10 days, until Yom Kippur, the day of atonement and the holiest day of the Jewish year, I wrote about all the reasons my life sucked. I documented my bad decisions, bad habits, and failed dreams and how I felt like a passive follower rather than an active protagonist in my own story.

What I hated most about my life was that I smoked cigarettes. So, on October 19, 1994, I made a decision: I would do a lifetime quit. Even if I didn't accomplish anything else, 1995 would still be a seminal year in my life—at least Mark 1996 would thank Mark 1994 for cutting out the cigarettes.

Throughout 1995, every day I didn't smoke was proof that I could control and actively direct my life. The next fall, I went back to my Book of Life. I got more ambitious with what I signed up for. That year I quit my job and launched my first company, FreeLoader. Let's acknowledge that it was both easier and harder than it would be today—easier because I had no opportunity cost and nothing great to walk away from, and harder because there was nothing great to turn to if the company failed. Amazingly, seven months later, FreeLoader was acquired for $38 million. That allowed me never to work for anyone again and enabled my career as a product maker and entrepreneur.

This experience became the foundation of my Book of Life practice for the next 30 years. Each year I ask: What would Future Mark thank Present Mark for doing? Then, I sign up for a specific change or goal and commit to making it happen.

Some years have been more successful than others. After 1994 and 1995, I thought I could do anything, so I committed to going on a date with Alicia Silverstone, who had just starred in *Clueless*. I never came close, even though a friend in LA told me she worked out at his gym. (I did eventually meet her in 2018 after we matched on Raya.)

When signing up for my next year, I like to focus on two types of actions or goals. The first is something in my control, like removing my wisdom teeth, or something I can change in my everyday habits, such as quitting alcohol, starting intermittent fasting, or having a daily step goal. Because these are in my control, I know that if I commit, I can succeed. Achieving these first actions helps me accomplish the much bigger second type of goal, which is usually a massive, life-changing dream, such as quitting my job and starting a new company.

My Book of Life hasn't just kept me accountable to personal goals; it's also kept me connected to the deeper "why" behind my work. Without this regular practice, it would be too easy to get distracted and lose sight of these convictions. For example, I'm a pretty good investor, but I will never be an amazing one. I don't care to become one and, honestly, can't see what difference it would ever make if I were. That's not my "why."

A Book of Life is like a compass that helps you come back to your North Star—what will matter to you over your entire life, not just today or even this year. It is what I credit for giving me the focus to found 10 companies and the discipline to sidestep countless dead ends.

Book of Life Practice

My Book of Life practice includes a specific book that I return to year after year. I write in it only during the 10 days between the Jewish High Holidays.

I start off on Rosh Hashanah by reading all of the past years' entries, cover to cover, to remember where I was, how I felt, and what I hoped for then. I write notes in the margins to my old self saying things like "Hey, good news, we did this" or "I still haven't done that."

I like to break the practice into three categories:

1. Taking stock of where I am at this moment. This is my "spiritual balance sheet."
2. Looking at the last year, how I did against my goals, which is my "spiritual income statement."
3. Notes on the next year and thinking about what I hope to get done. What can I commit to?

I begin each new year by writing about how I am feeling. I do that because when I read past years, I want to go back and *touch* that moment in time. I try to connect with Mark 1994. Do I get who he was and what he thought about? Do I know what he was worried about? As I look back, I see a lot of frustration that I hadn't committed to more and accomplished more.

Next, I note major events that happened in the past year. I devote a page to what stands out for me, good and bad: weddings, birthdays, health moments, major trips, relationships, challenges. At times, this page includes world events.

As I near the end of the 10 days, I start writing about what I can do in the coming year. For the past 20 years I've written, "Next year I'm going to launch Dot Earth." In 2006, I blogged that there should be an open data transfer protocol (like http) for more than just static web pages. I started dreaming of a metaverse, accessible through the web, that would allow us to connect to interactive experiences far beyond games. I imagined users typing in a single URL to accomplish something in the real world (e.g., I wanted to type zazie.colevalley.earth to get my favorite local coffee blend). I started calling this vision Dot Earth and imagined a new top-level domain system like Dot Com. It's an instinct that has haunted me for the past 20 years. It was one of the threads that led me to found Zynga. There, I returned to this idea so many times that my teams nicknamed it Mashed Potato Mountain, from the movie *Close Encounters of the Third Kind* (which you will get if you've seen it). In recent years, I've been working on a company called Erth.AI, which is building the first building block, a browser-based game engine that works in real time with large language models (LLMs). The goal is to enable anyone to vibe create interactive experiences that connect to the real world. I'm not the only one working on their vision for the metaverse. Zuck has invested over $80 billion and even changed his company name to Meta. Even though he is pulling back now, I'm still a believer.

Partner with Your Future Self

When I review my Book of Life each year, I'm having a conversation with myself across time. I look for patterns in how I've lived—the surf trips I repeat annually, ending the summers at Burning Man. Some years are indistinguishable, which is always a wake-up call, telling me it may be time for disruption.

As I look back, I see a painful pattern. The story my book seems to tell is that my biggest growth and company building moments have only come from my deepest despair. The more frustration I felt, the more likely that time was associated with me taking a disruptive action—usually starting a company. And the happier I have been with my life, the harder it has been for me to sign up for a big change. In 1994, when I was the most unhappy, I signed up for the biggest disruption of my life (quitting my job and basically burning my résumé) and achieved the biggest positive change, a lifelong career as a product founder. So, if you are feeling dissatisfied, the good news is that this is the best time to do something about that feeling.

Great product makers and entrepreneurs do something similar. Mark Zuckerberg signs up for one major change or goal every year and used to list these as "life events" on his Facebook timeline. Some more clearly involve partnering with future Zuck, such as learning Mandarin, while others are a bit less obvious, like eating what he killed for a year or wearing a tie every day in 2011. More recently, Zuck committed to becoming a mixed martial arts (MMA) fighter. Jeff Bezos has what he calls a regret minimization framework to help him make choices he'll be thankful for in the future.

The things we regret aren't the TV shows we missed or the extra day at the beach we skipped. We regret the book we never wrote, the instrument we never learned, the health issue we never addressed. We miss not taking the time to fix our teeth or our messed-up knee or running a marathon. I regret not learning guitar. I wish one of those "lost years" had been dedicated to that skill; I'd be playing today, thanking my past self. That's the point of partnering with your future self—making sure each year contains at least one meaningful accomplishment and that you are taking full advantage of this life.

In 2016, I lost my original Book of Life, which contained 22 years of reflections. While painful, this loss was an important reminder not to be too attached to anything. It also forced me to sit down and try to remember one seminal achievement from each year. For some years, I couldn't recall anything important. The blank spaces in that exercise taught me as much as the filled ones. That happened out of necessity, but it's a good exercise and a good place to start a Book of Life practice.

Build a Time Machine

I think of my Book of Life as a *time machine* that can take me backward and forward in my own life and help me make better decisions. I write notes to my past self in the margins—and many to my future self as well. It's a lifelong dialog. The practice of coming back to today can be a powerful tool to help motivate us to go after moonshot ideas but also to reset priorities and make hard decisions like killing marginal projects. This practice builds a hunger to disrupt my own patterns.

I saw Tony Robbins do this with 5,000 people in an odd but powerful exercise where he turned the lights off and asked us all to imagine our lives in 20 years if we never changed what we hated most about ourselves and our lives. For 30 minutes I listened as people wailed and cried.

For me this has always felt like a positive exercise (no wailing required).

34 years – My whole Adult life. so far.

JAN FEB MAR APR MAY JUN JUL AUG SEP OCT NOV DEC
1 2 3 4 5 6 7 8 9 10 11 12 13 14 15 16 17 18 19 20 21 22 23 24 25 26 27 28 29 30 31

Remembering Book of Life

- 2016 – Frank Z CEO. ~~Separate~~
- 2015 – CEO Z again.
- 2014 – Move to Aspen. Wyatt Born. Superlabs
- 2013 – Boy CEO of Z.
- 2012 – Z trouble. Hard year
- 2011 – Dad passes. Z IPO
- 2010 – Cityville. Carmen + Geo born.
- 2009 – Farmville. Zinga Passes.
- 2008 – Marries Ali.
- 2007 – Move to NYC. Start Z.
- 2006 – Tam passes July 25. E & C wedding. meet Ali
- 2005
- 2004 – Tribe. Invest in FB.
- 2003 – Tribe. Mo. First Bh. Ghetto Ranch Purchase.
- 2002 – Biking w/ Beerman + Vivant
- 2001 – SPRT IPO. Tank Hill w/ Martin.
- 2000 –
- 1999 – Jenny.
- 1998 – Starts Support.com. Buys Tahoe House.
- 1997 – Buys Shrader House. Tom + John.
- 1996 – FL Bought $38M. Moves to SF
- 1995 – Freeloader w/ Sunil
- 1994 – Columbia Capital. OC. Erin. 10/9/94 Quits Smoking + Start 4K
- 1993 – TCI. Denver.
- 1992 – HBS. Bain
- 1991 – HBS.
- 1990 – HK + Asia Travel. Asian Capital Partners. Marisel
- 1989 – Lazard. NYC w Rich + Bal
- 1988 – Graduate. Europe trip. Lazard
- 1987 – meet 4046 Friends
- 1986 – Fight w/ Dad. Wharton
- 1985 – U of M.
- 1984 – Graduate HS.

20 yrs ago.

If I come back in time from five years in the future, what will I thank myself for doing? Today, it's going all in on AI and launching products and companies that deliver on my vision of enabling everyone to live their lives at the speed of play.

Stop Time

There are moments when we realize history is happening around us. These moments, which can be externally driven or your own creation, are when I like to say out loud to myself and my teams, "We need to *stop time*."

This is a rallying cry to stop work as usual and get intense, often about a new direction. We have to check in on whether our path still makes sense given whatever new information we're seeing in the market, and perhaps make hard decisions like a small or big pivot. Then we need to be all in on sprinting to our, hopefully, more clear and relevant objective. Our teams often resist this; some may even opt out. But if you're right, the team will quickly show up in a much more committed way.

One powerful example was early in building Zynga. In the fall of 2007, the Facebook app ecosystem was exploding and so were our projects. Our teams were spread too thin across a bunch of word games and other small app ideas in addition to our core Zynga Poker game. I called everyone into a room one Monday and said we could keep on building these random new games like everyone else, maybe launching another 10, or double down on Poker, which was at 400,000 *Daily Active Users* (DAU), possibly doubling it to 800,000 and beyond. I saw the power and competitive advantage in building and running one "franchise" app versus treading water like everyone else, constantly launching new apps to replace the older ones nobody was supporting. It was a risky, uncomfortable bet with no data to prove this would work, and nobody liked it. I forced this decision through anyway, and after some grumbling and friction, we soon had the whole company focused on this one simple goal, which we did achieve. It was also instrumental in setting the strategy for the company, which is true to this day, of building *forever franchises* (we'll revisit the power of franchises later).

Time moves so fast that if we don't intentionally shape these blocks, we risk having nothing to show for them. Committing to a vision with specific outcomes is a way to stop time.

This practice spills over to being disruptive with your teams and products. It's not necessarily efficient. But being a great product leader and pursuing a big vision almost always means you will have to stop and switch to get the train moving in the right direction. These switches may happen frequently, because the right direction now is often very different from what it was last month or even last week. I always say to my teams, "We need to stop time, because this _____ is so important."

Stop Being an Expert Witness

My high school yearbook quote, which was written by my classmates, was "Some people have tact. Others tell the truth." This prophetic line would define my early career. When in business school and interviewing for Disney, I was asked how to bring Disney World to Middle America—a replica in every city. I told my interviewer what I thought: "That doesn't sound like a good strategy. Multiple lesser-quality parks will probably hurt your brand." I didn't get asked back for another interview.

Throughout my career, I found myself functioning as an *expert witness*—the person who is closest to the data (and the right answer) but furthest from the decision. During a summer internship at Bain & Company, I proudly presented evidence proving that in the snack food industry—the area I was investigating—Bain's foundational graph (which was created by Mitt Romney himself) was incorrect. Higher relative market shares did *not* always lead to higher return on sales. I thought Bain was going to offer me a full-time job on the spot. Instead, people were so offended that they walked out of my presentation. No one spoke to me for the rest of my internship.

Later, at TCI, which was then the biggest cable company in the world, run by the legendary John Malone, my boss asked me to analyze an opportunity to put $400 million into Prodigy, the biggest internet service of its time. "Prodigy is a bad service that loses massive money. There's this other newly public company, AOL, and we could buy that whole company for $110 million," I said. "Why don't we just do that?" TCI didn't do either deal. Prodigy ultimately went bankrupt. In another meeting, after I presented to Malone why we shouldn't do a deal he really wanted to get done, he said, "I don't need some wet-behind-the-ears MBA telling me what's a good investment." That seemed like the end of my career at TCI.

In each of these jobs, I was the victim in a storyline that kept repeating: The adults would make the final decision, and the underlings like me would then go try to make it work—and try to limit the damage. I didn't believe in paying dues and working my way up corporate ladders. I struggled with the management principle of "disagree and commit." I was more "disagree and disagree." I didn't make a good employee, although I later realized I was exactly the kind of hire I wanted on my future teams.

The expert witness has been an important concept throughout my career. I was a frustrated expert witness in jobs I had in my twenties; later, I learned to hire other expert witnesses and set them free as "CEOs" by giving them the authority to make decisions in their area of expertise. What's your current situation? Are you an expert witness, close to the answers but far from the authority to implement them? Or are you in a founder position, and perhaps you can hire expert witnesses and set them loose to prove they're right?

We All Start as Outsiders

My instinct for my first company, FreeLoader, started in the middle of the night in 1994. I was haunted by this feeling that my life was going nowhere. At the same time this thing I'd been waiting for and dreaming about—this public online network (the internet)—was happening right in front of me, and I wasn't a part of it. I was filled with FOMO. I felt like an outsider.

Ten years earlier, when I was in college, my brain was infected by George Gilder's book *Microcosm*. He pointed out that the previous 100 years had been about the macrocosm, or the physical material world, where fortunes were made from laying railroads and building shopping malls. The next 100 years would be all about the microcosm, or the microchips connected to networks, spreading ideas and knowledge at the speed of light. He predicted that, just as calculators went from taking up an entire room in the 1950s to becoming keychains given away at the bank, everything around us would eventually be infused with productivity gains, becoming abundant and almost disposable.

Now, a decade later, everything was happening in Silicon Valley and Seattle. I found myself trapped in DC, which felt like the furthest possible place from this tech future. Even though I was such an outsider, I felt at a deep instinctive level that I could visualize the whole thing. But I didn't know how to code. I hadn't worked at any tech companies. I had no credibility.

The only thing I could think to do was write a long essay about what was happening. I stayed up all night writing about how the Mosaic/Netscape browser and the publicly available internet were going to end the Microsoft monopoly. The browser was going to break open software and internet and consumer services so that any entrepreneur could compete.

I emailed my essay to the editor of *Interactive Week*, the only publication that was covering new media at the time. I had nothing to lose and not that much to gain. I just had to tell someone what I had figured out and to see whether they thought I was right.

The editor emailed right back asking, "Who are you?" Then he used this thesis as a front-page story under his byline without mentioning me. I didn't care. I was impressed that *Interactive Week* thought that I—a nobody, an outsider—had figured things out to the point that they wanted to publish my essay as their own.

The editor threw me a bone when a week or two later he interviewed me. He published a back-of-the-book profile on me as an up-and-coming venture capitalist. Fred Wilson, a junior partner at Euclid Partners, a little venture firm in New York, read the piece and got in touch. Fred was the lone partner at his firm who wanted to do software and internet. When he called, I thought he was interviewing me to be an associate. But when I got there, I saw that his office was so tiny and so full of papers and books that there was no way he could give me a job; there was literally nowhere to put me.

"The world doesn't need another VC," Fred said. "What we really need are entrepreneurs we can back. Do you have any ideas?"

I wanted to build consumer software for the web. That was exactly what I had written about in my Book of Life. I had committed to quit my job and go for it, even if at the time I didn't know what that looked like. So when Fred gave me this small entry, I pounced on it. And it eventually became FreeLoader.

Build a House You Want to Live In

Part of what I get from my Book of Life practice is clarity about my intentions. Each time I've started a new company, I've tried not to make the same mistakes, though sometimes I've overcompensated and made new ones.

By the time I founded Zynga at 41, I had started three other companies—one acquired, one public, and one failed—and learned hard lessons about control. It wasn't fashionable to still be founding companies at that age. Many of my founder peers had become VCs, and conventional wisdom claimed backing founders over 30 was a mistake. But I wasn't doing this for a job or money; I was doing this for my passion and satisfaction.

In all my previous companies, especially Support.com, I made so many compromises that paradoxically, as the company achieved bigger milestones, it became less and less a place where I wanted to work. ("Know your goal, or suffer death by 1,000 compromises.") These compromises involved endless small decisions about location, office space, board and team members. Many founders wake up one day and say, "This isn't fun anymore, but look what I achieved. I guess I'm just a casualty of our success." Then it becomes acceptable to think that the tour of duty's over, and it's time to hire a CEO. VCs will never argue with that decision; they love it. But it's rarely the right long-term choice for the company.

As founders, we make sacrifices, similar to parents. We put the needs of the mission and organization ahead of our own. But if founders are the most essential employees—the keepers of the flame for the mission, vision, and cultural DNA—how can a company afford to lose them and still remain special?

As I built Zynga, my mantra was "Build a house you want to live in." The mantra came directly from my Book of Life reflections, helping me stay connected to my "why" and reminding me why control matters. Instead of thinking about everyone else, I focused on creating a company and environment that made me excited.

That's why I purchased and remodeled an old potato chip factory a year before founding Zynga. Financially, this made no sense. I put millions into buying and renovating the former Williams & Company building, and then we raised millions for Zynga's Series A. We called the building the Chip Factory, a double entendre, given that it went from creating potato chips to creating poker chips. The building may be worth twice as much today, 18 years later, while those same dollars in Zynga became worth billions.

Too often, we underestimate how our workspace affects our creative success. I remember hearing that Steve Jobs insisted on approving carpet choices for conference rooms; this fact was cited as evidence of founders' insane micromanagement. I see it as beautiful: He cared so much about the product that his work environment needed to be perfect, too.

I didn't go that far, but I insisted on a commercial kitchen with trained chefs (students from the culinary school across the street). At Tribe, we ate pizza in a dusty warehouse. Back then, I was always excited to visit our VCs because they had beautiful catered lunches. With Zynga, I wanted VCs to visit and think, "They live better than we do!"

Since we were working constantly, the least we could do was live well. In addition to great food, the people in each game studio designed their own space. Even with 1,200 people working there, it felt like a federation of start-ups: Mafia Wars had a meditation room; YoVille had a video game room.

My Founder Mode

My friend Brian Chesky, the Airbnb cofounder and CEO, inspired the term *founder mode* during a 2024 talk at Y Combinator (YC), the famous incubator that has fostered other massive companies like Stripe and Dropbox. I define founder mode as unapologetic leadership—having the conviction to lead through chaos, even when you're losing and it's not obvious your instincts are right.

Most start-ups hit moments when the original idea *isn't working*. That's when the board gets nervous, the team looks to you for confidence, and everyone wants an easy, safe answer. But great companies aren't built on safe bets; they're built on bold, uncomfortable decisions that might make you unpopular in the short term but that drive long-term success. Are you ready to be disliked or to stand alone if necessary?

Too often, we over-tune to stakeholders and under-tune to our instincts. Founder mode gives us permission to be our true selves—not to be jerks, but to take the wheel and ignore consensus when necessary. I've always referred to my companies as "democratic dictatorships": Everyone can voice an opinion, and then I make the final decision. Lately, we have seen bold CEOs at Palantir, Coinbase, and even Google tell employees they can leave if they don't agree with their positions. Start-ups aren't democracies.

No one will grant you founder mode. Most investors believe 99 percent of founders aren't qualified for it. Some say founder mode is just a way to give bad founders an excuse to be bad. VCs will always look back and say, "Of course Bezos should have had founder control" (he didn't). Or Elon. But nobody thought founders should have control at their low points when they wanted to do things everyone else thought were stupid.

Founder mode isn't about being reckless; it's having the control and conviction to act without asking permission. Without it, you're just an employee waiting to get fired from your own company.

Always Keep Control

My father hammered the idea of control into my head. I failed to take his advice throughout the course of my first three companies.

I finally got the point by the time I was raising money for Zynga. I was up front about maintaining control and half jokingly said it was to protect investors from themselves. I started investor calls with "Here are the top 10 reasons you won't want to invest in my company. If you still want to talk, great." Ninety-five percent said, "Thanks for saving me time," and hung up. Many Valley investors have stories about "missing" Zynga. The truth is they turned us down because they wouldn't accept my terms.

My approach made fundraising harder. Zynga should have been the easiest round ever—we had $200K in monthly free cash flow growing at 30 percent, and I had a proven track record. Yet I struggled to raise $5 million and took a $15 million pre-money valuation. Fred Wilson made me eat the entire option pool. Today, the average Y Combinator company gets better terms than I did then.

When Sequoia questioned our valuation, asking how I could justify $20 million pre-money versus $15 million, I said, "I'm going to build a multibillion-dollar company or fail. If I succeed, it'll be worth a lot; if I fail, it won't be worth anything. If you're worried about the difference between $15 and $20 million, this isn't the right investment." We're seeing ever higher valuations today largely for this reason. VCs today mostly get this trade-off and view most seed and Series A investments as call options on big outcomes.

Sometimes you have to go slow to go fast. Ultimately, I created conditions for success by refusing to work for a VC-controlled company. I told investors we were mutually choosing each other; they couldn't just name board members. The board is a company's DNA, and I wouldn't accept a junior partner and the resulting "junior partner syndrome," in which I'd have to negotiate with an entire VC firm through someone who was only an expert witness at their own firm. We would never sell unless we failed.

The best VCs (many of whom are former founders) know how to support founders while getting out of the way. And world-class VCs, including Fred Wilson, Reid Hoffman, and John Doerr, know to back founders in their darkest moments. I was lucky to have them at Zynga.

The Abyss

One thing most founders share is the experience of being in the Abyss. For most of us, this place before and after our start-ups or even jobs we have loved is open-ended and dark. It's unstructured. The world doesn't care whether we get out of bed in the morning. We lose purpose and don't know whether, how, or when we will ever reemerge.

The Abyss also has value that we're unaware of when we're in this state, however. Once we get over our anxiety about being in the Abyss, it becomes an opportunity to explore our curiosities. Giving ourselves permission to pursue things that are not obviously useful or productive is critical. These explorations help us tune into our instincts, hone our ideas, and develop our *taste*.

Support.com, my second company, went public during the last week of the dot-com bubble. Within six months, everything crashed. By March 2001, San Francisco and the consumer internet had entered what I called nuclear winter. South of Market Street, which had been overrun with start-ups, became eerily empty—like a ghost town after the end of the gold rush. Everyone I knew left, and many moved to Tahoe, where real estate prices soared.

That was my first period of unemployment. Tom Cole (who, by then, had been living on my couch for eight years) and I retreated to my house in Cole Valley. I went through a year and a half in the Abyss. I spent my time wandering the neighborhood with my dog Zinga, smoking pot, and meditating. Tom was running a kief bar in my basement (this was before the advent of legal dispensaries). He was a pot mixologist (far ahead of his time), mashing different strains into powders, drying them in martini glasses on my back porch, and putting the concoctions in little labeled bottles with names like "bubblegum." A bunch of ex-FreeLoader people and other Haight-Ashbury locals would show up at the house to smoke at Tom's bar.

During this time, I started developing what the legendary business coach Bill Campbell would later call my "18-month thinking." He said, "You're one of the best 18-month thinkers I've ever met. You can see around corners. Trust that." What Bill meant was that I had a keen sense of what people would want in the near future. My own take has always been that my usage patterns represent what the early majority of the mass market will want.

While everyone else was declaring the internet dead, I tried to stay close to first principles. I was paying attention to consumer adoption, which never stopped.

Because I wasn't doing anything and companies weren't getting funded anyway, I had the luxury to just think about the future. I started to feel a deep well of desire to be engaged in something—almost anything—that would feel productive.

One great thing we can all do in the Abyss is sample everything. I went deeper into cooking and particularly my obsession with the perfect Bolognese sauce. I went to Italy with my girlfriend, Jenny. I decided to eat only pasta with Bolognese at every meal, in my search for the perfect Bolognese. That, and my desire to drive like an Italian race car driver, made Jenny want to break up with me daily. But my search wasn't just about the sauce; it was about developing the ability to recognize when something truly works.

During this time, I developed a deeper theory of the future cultural revolution that would be set off by the worldwide internet. I called it the "Revolution of the Ants." I believed that once everyone had unlimited access to this network, we the people would rise up, self-aggregate, and eradicate the need for the gatekeepers—the brokers, the middlemen—

whether politicians or newspapers. Money and power would no longer be derived from controlling the message or our access to one another. Instead, there would be a perfect market in which the best ideas would spread. A few years later, I captured this theory in a blog post I wrote on my Blackberry while stuck on a plane. It represented a deep instinct well that would eventually be the people web (social media and social networking).

A small group of us still held a flame for the consumer internet during this nuclear winter. Reid Hoffman and I started to meet with others in San Francisco, talking about Web 2.0. I hosted brainstorming sessions at my house in Cole Valley, which I had transformed into a three-story one-bedroom loft (not a smart real estate move!). I covered the walls with big sticky sheets from these sessions, capturing what we saw as the instinctive pillars of the next wave.

We all realized a core rule of the internet was that "anything that can be free will be," and our user data would be set free, too. We saw a future web made up of people, not pages. While Google had built its empire on page rank relevancy, we believed the next wave would be built on people relevance. Reid had the idea for LinkedIn. I had the idea for Tribe.

The Abyss is often where our deepest instincts form. It's where we start seeing patterns that others miss because we're forced to slow down and really look. Many founders rush through this phase, desperate to get back to "productive" work. But without my time in the Abyss, there would have been no Zynga, no understanding of the social web, no clarity about what I wanted to build.

Imagination Chessboard

In 2003, when I was 38, I started working with a life coach named Erika. I wanted two things: a life partner and a life mission. I'd already had success beyond my hopes and dreams, but I wasn't satisfied.

Tony Robbins talks about how happiness is where your life meets or exceeds your expectations. By that measure, I should have been happy. But I felt a deeper void. I was happy day-to-day with success, friends, and fun, but my life lacked meaning.

Erika didn't let me off easy. In our first session, she said, "You're emotionally unavailable. You're not going to find anyone because you don't actually want to." She told me emotionally unavailable people are like alcoholics who can find one another across a bar.

Erika helped me develop what I call my *imagination chessboard*, which helped me envision what I wanted my life to look like two or five years out. In practice I literally wrote out whatever story I wanted my next life chapter to be. In the Netflix show *The Queen's Gambit*, the young chess prodigy lay in bed at night and imagined the winning chessboard and the different moves that would get her there. If we can picture it in our mind's eye, we can get there. We know our North Star.

On the personal side, instead of helping me get more dates, she put me on a six-month dating moratorium. For the first time in my life, I wasn't trying to find someone. Instead, I was clearing out old cobwebs; she literally made me clean out my closets. I had to examine behaviors that didn't line up with my stated objectives.

On the professional side, my imagination chessboard helped reveal two limiting beliefs that were holding me back from being creatively productive. The first was that I needed my own capital—a thread throughout my whole career and probably true for most other founders, too. I thought if I had $500 million, I could stop worrying about funding and play the long game, build things right, and go for bigger, more creative ideas instead of just what could get VC funding or immediate revenue. The second was my perception that I needed a permanent incubator. I dreamed of having an idea factory—what Bill Gross had done with Idealab, where he raised a large pool of capital and spun out new ventures based on his many ideas.

This exercise showed me that I have often been stuck because of my own limiting beliefs. And if I could get past these, I could start to go deeper on my why. I realized this wasn't about needing money or having a platform. It was about using 99 percent of my full creative capacity.

In building Zynga, I used this imagination chessboard to get my teams past their own limiting beliefs. I often challenged them to answer, "What if everything goes right?"

My best use of this was in 2007, when EA called me in to meet with their lawyers, whom they conveniently called "business affairs" to sound less scary. Zynga had built a series of games that resembled popular Hasbro titles like Boggle, and EA had the rights, so they threatened to sue us. Erika challenged me to imagine what the best possible meeting would be. I said, "Zynga should be EA's new Walmart"—meaning their online distribution via social networks to the mass market. So that's the pitch I prepared. Luckily for me, Bing Gordon decided to drop into the meeting. When he heard me say the Walmart line, he took over the meeting and the whiteboard. EA decided not to sue, and Bing ended up helping me build Zynga.

Today AI can help you develop your imagination chessboard in minutes. Close your eyes and imagine life as it will be in the future when you have achieved your North Star. I create these chessboards all the time. I walk around having conversations with ChatGPT and use each imagination chessboard to go deeper on product use cases. I prompt it to write about life in two years, when I have built my vision for the metaverse and millions of people rely on my platform Erth.AI to turn their ideas into realities for others to consume and love. It helps me live my life at the speed of play and move from a high-level vision to detailed use cases that I can start to play with, test, and iterate on faster.

TL;DR

Keeping a Book of Life practice helps you stay connected to your North Star across time. By partnering with your future self, you create accountability to goals that truly matter—not this week's distractions, but what you'll thank yourself for years from now. The exercise is simple: Each year, take stock of where you are, review how you did against last year's goals, and commit to what you'll accomplish next. Without this regular practice, it's too easy to drift through life as a passive observer rather than an active protagonist in your own story. What could make this or next year a seminal year in your life?

2

Instincts Versus Ideas

We're all guilty of falling in love with our own ideas. This is the number one cause of death of consumer start-ups. Unfortunately, enough false positives can keep the funding and effort alive for a while. But it's amazing how many founders get to their big idea once they run out of money. Why? They're forced to get real. Once failure isn't an option, we move faster, ruthlessly kill ideas, and become heat-seeking missiles.

We fall in love with our ideas because they're based on instincts that are right, probably even more right than we realize. These instincts are massive hidden icebergs. So, yes, you're onto something big. But the idea variant you're in love with isn't likely to be it.

What if you knew your *instincts* were always right but your *ideas* were usually wrong? Would you prosecute your ideas differently? This notion has been a fundamental building block of my career as a product maker. I've found that for good entrepreneurs, our instincts are right 95 percent of the time, but our ideas are right at best 25 percent of the time.

We've all had that moment when we've seen a new product and thought, "I had that idea!" The truth is, we didn't really have *that* idea; we had the core instinct it was built on.

An instinct is something we feel in our gut. It's a connection at a deeper human level, often the feeling that something is missing or could be far better. Take the iPhone: Every one of us, somewhere inside, knew there was no reason a phone and computer couldn't merge into one device. That was the instinct. The idea—the specific instantiation—was a product only Steve Jobs could deliver.

In 2003, I had the instinct that there must be a better way to get a taxi than calling Luxor Cab and hearing "1 minute" before the dispatcher hung up. My idea was to order a taxi by text messaging the dispatcher. I even bought the domain smstaxi.com. The instinct was right, but this was never going to be Uber. The idea wasn't there yet.

Six years later, Travis Kalanick and Garrett Camp had the winning idea, enabled by iPhones, GPS, and the brilliant insight that we could trust non-professional strangers to be our drivers.

Throughout my career, I've pursued what I call *instinct veins*—deep sources of insight about human needs and behavior that can spawn multiple product ideas, even whole new industries. These veins are so fundamental that you can return to them again and again, each time extracting new possibilities. Two of my most powerful instincts have been "It Just Works," which was the foundation of FreeLoader, Support.com, and Brightmail (which my FreeLoader cofounder Sunil created), and the *Cocktail Party*, which led me to found Tribe.net and Zynga and to make seed investments in Facebook and Twitter. (And to show just how much "liquid gold" lies under some of these veins: My $38K investment in Facebook—which purchased nine million shares, or just about 0.5 percent of the company—would be worth around $6 billion today if I'd had the foresight to hold the shares.)

My First Instinct Vein: It Just Works

During most of my time as a VC at Columbia Capital, I was obsessed with the web and the consumer internet, with the burning feeling that I should be starting something. My friend Brad Zions was sick of hearing me talk about it and suggested I meet Sunil Paul, who worked with him at AOL. Sunil had been a rocket engineer at NASA and was AOL's first (and at the time, only) internet product manager. He was responsible for creating

AOL's early internet capabilities, and like me, he had no one to talk to about everything that was going on.

Sunil and I met at a Pizzeria Uno, and I immediately saw that he could be the perfect cofounder. He had the credibility I lacked and an idea to develop an "internet box" that would make web access simple for everyone. He literally wanted to build and sell a box with a phone modem and software to dial into internet service providers such as AOL built in. I also wanted to make the web more accessible, but I wanted to offer a software-only solution, not hardware, which sounded like a heavy, slow, expensive endeavor. I believed software could be distributed for free to millions of people on the internet while hardware would have to move (slowly) through retail channels. We wrestled over which path to take. Mine won. We pursued a software-first approach because it would require less money and less time to prove itself.

Although Sunil and I initially had different ideas, we had the same instinct, that PCs and the web should "just work." In fact, we started to refer to whatever we were making as "It Just Works (IJW)," and I registered IJW.com. (This was so early that three-letter domains were available for $20.)

Our instinct—that the internet was too slow and too hard to navigate for mass-market consumers—was inspired by what we saw happening with Mosaic, the first popular graphical web browser. Mosaic, which launched in 1993 and could be downloaded for free, was still primarily for hobbyists and was too hard for most people to use. It was not obvious yet that Mosaic (and, soon, Netscape) represented the beginning of the modern web, which presented a landscape wide open with opportunity. I believed that there was going to be a world of consumer internet applications that made new services accessible to the mass market. I also had the instinct that consumers would massively adopt free software.

We set out to build a free product that would offer users a toolbar, attached to their Netscape browser, allowing them to subscribe to any website they chose. Our service would dial the internet in the middle of the night and download everything that was interesting—stocks, news, sports, movie trailers—and then display it throughout the day as an interactive screensaver with ads.

Founding FreeLoader: Funding Your Instincts

Here's the long and crazy story about how I founded my first company, FreeLoader. There were a lot of lessons for me about instincts, right partners, and control.

As cofounders, Sunil and I each needed to put up $30K to seed the company, since we had no funding. I quit Columbia Capital with $80K in savings. Around this time, I shorted the three publicly traded internet service providers (ISPs)—NetCom, PSI, and UUNET—in the belief that they were commodity services that would ultimately have no margins. All three went straight up from the day I shorted them, and my bank account quickly went to zero. (Eventually all three ISPs went bankrupt. Life lesson: Never short stocks.)

My best friend Tom Cole had inherited $100K years earlier, which he was living off (mainly on my couch). Luckily, he agreed to loan me $22K, and I agreed to give him a job. I put the remaining $8K on Best Buy and Target credit cards, which offered 14 months of credit interest-free.

In October 1995, Fred Wilson, still a junior partner at Euclid Partners, agreed to lend us $250K with the God-fearing term of a four-month demand note. We would have to get a venture deal done by the end of February 1996 or shut down.

We started calling the company OffLoader. Then one day, Sunil looked at Tom (who was still living on my couch) and said, "FreeLoader." We registered freeloader.com, since most common English words were still available. When we asked people what they thought of the name, they said they hated it. But they also remembered it. That's when I started learning about inventing brand names on the internet. Weird names that hit emotional chords are good, because people remember them enough to type them into a browser or Google. It's even more important if you're hoping to generate word of mouth. This early instinct about naming stuck. It's also the reason I've always believed that using actual words is powerful: They're already in our brains, and people can remember them more easily. Conversely, a weird new name with weird spelling, such as Zynga, is harder for people to remember at first but can be stronger later if the name breaks through. I always chuckle when people say Zynga is such a Web 2.0 name. The reality is that many entrepreneurs during that time were driven to creative spellings due to lack of available domain names.

Every time you start a new product, the world tells you, "It already exists, and we don't need it." I always wonder how both can be true. In the case of FreeLoader, the product did already exist. Microsoft and Netscape offered what they called offline browsing. Berkeley Systems was generating $30 million a year selling static screensavers with flying toasters for $35 at Best Buy. We were offering a screensaver with dynamically updated interactive content for free. This was a classic disruption—if our ad business could work.

In January 1996, a few months into our product development, I found out that another service called PointCast was launching. PointCast was built out of the same instinct but was far better resourced and further along in development with a slick, polished product. FreeLoader was an open system allowing users access to any content on the web. PointCast was more like AOL, which offered 8 to 10 preset content channels and no ability for users to choose their own content. Both AOL and PointCast were walled gardens—closed, carefully managed, beautiful experiences. Both lacked the infinite, open nature of the web.

I knew we were in trouble when I ran into an old friend at one of the first internet conferences and she shared the PointCast advertiser brochure with screenshots. PointCast had built its software "correctly" over a much longer time. It was full of easy-to-access content covering most of what anyone wanted, from stock quotes to weather to sports. We had no product, no money, and no marketing, but we did have a belief that ultimately our open web approach would beat PointCast's polished, closed approach.

We were in an incredible race against time. Every day was a sprint to beat PointCast to market before we ran out of money. Sleeping on couches in the office. Living on caffeine. The only lever I had to get ahead of PointCast was PR. We broke their story just to position ourselves as the number two in this important new space, which became known as push technology.

This is where I first learned a core founder approach to time cycles:

A day is a week.

A week is a month.

A month is a year.

Every week, we put out a press release announcing new content and distribution partnerships with sites such as Yahoo, Excite, and ZDNet. When users downloaded FreeLoader from those sites, they would subscribe to daily content updates. For some of these sites, we eventually generated 10 percent of total page views.

There was no ad-supported business model on the internet yet. CNET was the one company trying to prove the ad model. I remember CNET, in a ballsy move, buying out the whole internet in late 1995 for $1 million. Every page you loaded for a few weeks had a huge CNET banner, which did brand the company forever. It turned out to be a smart play. It generally costs 100 times less to buy awareness at the onset of new platforms, before there are established businesses and clear customer value. Note to all of us: Bet big on mindshare for emergent consumer platforms when they're cheap and early in their adoption cycle.

FreeLoader was wildly risky and not what traditional VCs were looking for. Imagine the pitch: *We have an unproven business model, unproven distribution, and an unproven product. Want to invest?!* By February 1996, we had no funding prospects and no money left. Fred's demand note was about to be due, and that would be the end for us.

Then, one day in early February 1996, we got a call from Jerry Colonna at one of our distribution partners, CMGI, which owned the search engine Lycos. The CEO, David Wetherell, was interested in investing.

CMGI moved quickly and delivered an exciting term sheet for $5 million (beyond the $3 million we were raising). The catch, which wasn't small, was that half the money would be given up front, and the other half would arrive once we met CMGI's conditions. One condition was hiring a CEO that Wetherell liked. Another was that we had to end our partnerships with Yahoo, Infoseek, and Excite, making Lycos our sole distributor. We'd basically be a captive company to Wetherell and CMGI.

I got Wetherell on the phone with Fred. "David, what's the deal with us having partnerships with Lycos versus Yahoo and everyone else?"

Wetherell had a big temper and lost it, perfectly on cue. "There's no fucking way you're working with anyone other than us," he said.

"Then there's no fucking way you're investing in FreeLoader," Fred replied without missing a beat. It was one of the most heroic moments I've witnessed from a VC. It cemented a lifelong friendship and partnership. (Jerry Colonna ultimately left CMGI and joined Fred in founding Flatiron Partners, which is today Union Square Ventures, a very successful VC firm.) It also offered a lesson that would come back again and again: the importance of keeping control.

We had no funding and no money left, but then, in the middle of February 1996, PointCast launched and changed everything. We got a call from Eric Hippeau's office at SoftBank. He was the managing director of the fund at the time. Masa Son, who founded and ran SoftBank, was always the first to believe and the biggest to bet on the next platform. He had put up $100 million for a third of Yahoo. SoftBank had bought ZDNet, and its team, headed by Eric, was tasked with investing in every internet category.

Eric told me that SoftBank had put push technology on the office whiteboard. The hype around the PointCast experience was so great that many started to fear that push technology would change the entire way internet users consumed content. All content would be "pushed" to users' desktops, and the "pull"-based method of slowly loading web pages would go away. All the hype was only because of the explosive PointCast launch. But now, with PointCast on every desktop, all anyone could talk about was push technology.

SoftBank, along with Fred's firm, Euclid, gave us $3 million. With that, we were able to continue. We got FreeLoader out in April. By today's standard of "if you're not embarrassed, you're not shipping early enough," we were on the bleeding edge. Our app crashed often, and our toolbar was based on a hack of the Netscape browser and Windows operating system that let us redraw our user experience (UX) below the browser window. The net effect was that if you ever moved or resized your browser, our toolbar might get lost somewhere else on your screen. But when it worked, it delivered a magical experience. We got two million installs in our first month!

PointCast had already built user interest and turned the whole category of push technology into one of the first consumer internet memes. Fox / News Corp tried to buy PointCast for $450 million, which the founder, Christopher Hassett, turned down. We were the next best option. Sometimes it's better to be number two. You're not giving up the world selling early. In fact, there are many cases of a number two monetizing while a leader loses.

Luckily for us, even if we were a distant number two, we had hit on an instinct that was right. As a result, leading players had to own this technology, and two of the then biggest public internet companies were interested in buying us. Individual Inc. offered $38 million, and Yahoo discussed an offer at $45 million. I took the Individual deal because I assumed I'd get fired by Yahoo in the first year. (Yahoo had just gone public and was only able to do a pooling of equity purchase, which would have preserved my four-year vesting that was in place. Individual did a purchase accounting transaction, which enabled us to fully accelerate our vesting, so we weren't at risk of being fired and losing our equity. Yahoo's offer risked leaving me with a quarter of my equity as opposed to a guaranteed deal with Individual.) In hindsight, the deal was a gigantic miss. Individual's stock price went from $35 to $6. Yahoo's market cap went from around $800 million in 1996 to over $100 billion at the height of the dot-com bubble. Any amount of Yahoo stock would've been massive. Still, we sold to Individual for $38 million seven months after we founded the company, making it the best performing investment in the portfolios of Euclid and SoftBank at that time. It all happened so fast that I hadn't even paid back the credit cards I'd used to get started. PointCast, after turning down $500 million, ended at zero, because push technology was over in a year.

That sale put me on a lifetime career path as an entrepreneur and product maker. I went from having no money and putting everything on credit cards to netting a few million dollars. I made a list of everything I was going to buy now that I was *rich*! The list totaled $9K; that included a new couch, a new TV, and a leather interior for my Nissan Pathfinder (which I still own).

To this day, it's not lost on me how thin the line is between huge success and total failure, and how unlikely this outcome was and how easily it could have turned out differently for me. The reality is that push technology became irrelevant within a year as the world moved to faster bandwidth connections. If FreeLoader had continued on, it would have failed, and I don't know where that would have left me.

Selling the company, in that moment, was a no-brainer. It was more money than I had ever imagined. It was not lost on me that I had gone from a small life where I was out of position, missing the internet wave, and had no pathway to success, to now being party to one of the biggest internet transactions of the year. And, it allowed me to lock in a win. As is the case for so many first-time entrepreneurs, if and when you get that chance to sell, it's nearly impossible to say no. Evan Williams sold his first company, Pyra Labs, which pioneered blogging, to Google for an estimated $20 million. That enabled him to go on to found something much bigger later, which became Twitter.

The FreeLoader experience gave me deep empathy for all founders. I can feel in my body to this day that nightmarish feeling of building products and companies, and being on that start-up path. It's an exciting adventure, but you never get a real night of sleep, because you are living on borrowed time. Yet my cofounder Sunil Paul and the rest of the FreeLoader team remember those experiences with love and admiration. This is true for all the teams I've worked with. We challenged and stretched ourselves in the space of months, and that's what makes us keep coming back to it. We're addicted to the challenge and the growth.

The world looks backward at the money made in the small number of success cases and rewrites their stories to say that founders did it for the money. However, as founders, we know that is not true, because it was so unlikely and so hard to get to success. If we were only thinking about expected returns, we all would have given up.

FreeLoader Epilogue

Right after the FreeLoader acquisition, Sunil and I moved to San Francisco. My old friend Tom and my new dog Zinga drove me across the country. We gave Tom the title "Chief FreeLoader," a role that was really a combination of mascot and office manager. He mainly went to Staples and smoked a lot of pot, which he also supplied to the whole company. (Unfortunately, Tom passed away suddenly a few years later on Zinga's tenth birthday. Today, their ashes are buried next to each other in a shrine on Aspen Mountain.)

A few weeks after acquiring FreeLoader, the CEO of Individual had a nervous breakdown. We were told by people in the corporate office that he tore off his clothes and ran around the office naked. He left shortly afterward, and the stock tanked to roughly $6 per share. The company became what I later referred to as a ***ghost ship***—a founderless public company run by a professional board that lacked any ownership. "Founderless" almost always means rudderless.

Eventually Individual hired a new CEO out of McKinsey. He was the Antichrist of business, with one of those fail-upward résumés. Sunil and I were considered kids, so they froze us, telling us we couldn't grow or shrink FreeLoader. They basically forced us to drink lattes for the next year until they shut our product/service down.

Choose the Right Body of Water

The body of water we chose for FreeLoader—free consumer internet services—is so big that today, we don't even consider it one industry. This wasn't obvious in 1995, when retail software was rumored to have a total addressable market of $35 billion, versus free internet services at $100 million. Retail software turned out to be a terrible market that dried up quickly, declining 90 percent over the next 7 to 10 years to $3 billion.

If you choose the right body of water, most boats will float, and if you choose the wrong one, even the best will sit on a dry bed. By "body of water," I mean the industry or specific market you want to address. Having a winning instinct isn't enough; it's crucial that you productize it against a huge (and potentially growing) market. That market may not be obvious when you start, as was the case for FreeLoader, but picking the right one means you're likely to succeed even if your idea isn't worthy of an A. Picking the wrong market means you'll fail, even with an amazing idea.

For FreeLoader, I pursued a deep instinct in the right body of water with a B+ idea. When we had focus groups test the product, people's responses were mostly "meh." The instinct to make the web easier to navigate was a deep vein that I would tap again and again, and our business plan—to sell ads to reach what we hoped would be a large audience—was unproven in 1995. We launched a freemium service because we lacked the resources to build anything else, and maybe by luck, we stumbled into what became a massive ocean.

The good news is that today, we can all be much more analytical in picking the right body of water. With Support.com, as I will share later, I forced myself and the team to connect the dots and pick a clear market with long-term growth potential. Often, picking the right body of water can be contrarian to what is hot in the moment (and therefore VC fundable).

My best friend Gary Leff is a talented entrepreneur and built Stir Crazy into a beloved chain of healthy Asian stir-fry restaurants. But his amazing boat was trapped in the dry bed of all oceans, the restaurant industry. After 10 years of building and running nine restaurants across the country, all profitable, he sold the company for an amount slightly greater than the capital invested. Word of advice: Don't start or invest in a restaurant, even if the restaurant will be run by your best friend and they are a rock star.

If you're left wondering how you will ever pick your right market, here are some quick criteria. (Sophisticated founders can skip this list!)

1. **Market entry potential.** Available and not crowded? Hot categories, such as AI coding agents today, are often so competitive that it becomes impossible to stand out or keep up.
2. **Proven business.** There is a clear line of sight to revenue and sustainable profits.
3. **Opportunity to innovate.** I love mature markets that are actually growth markets waiting for innovation.
4. **Long-term growth potential.** The market will benefit from macro trends you believe in.

Kill Your Losing Ideas

The path to founding Support.com was a lesson in the power of killing your losing ideas.

With nothing to do in the summer of 1996, Sunil and I started going out to lunch every day and brainstorming. We came back to our core instinct vein of It Just Works, which had led us to found FreeLoader.

FreeLoader was actually the first part of a much bigger business plan that we called PC-One. Our concept was to offer people a free PC with a $20-per-month internet service with a managed experience and ads. Our vision was to become an easy on-ramp for the mass market to the internet, the safe way to move from your AOL to a much more vast experience.

I didn't think anyone would say no to a free PC, but we needed to see whether there was any *heat* beyond the free aspect. Sunil and I stood outside a Tower Records (RIP) store in San Francisco and gave people $5 gift certificates to talk to us. We quickly learned that nobody trusted the idea of "free." Worse, people were afraid of new PCs. The prospect of reinstalling their software apps and kids' games was overwhelming. Also, the thing people hated most about their internet experience was email spam.

We quickly found that PC-One was a classic example of a winning instinct trapped inside a losing idea. But we also had surfaced two new ideas from our user interviews. With that, I did what has since become a defining practice of my career: I killed my losing idea. The free internet PC idea was dead, at least to us. A guy we had recruited to be the CEO pursued this idea without us and founded PeoplePC, which raised $65 million, went public, eventually declined, and ultimately failed as a service (because it was a bad idea).

Sunil and I dug deeper into the It Just Works vein. Sunil turned this instinct into Brightmail, a consumer spam-blocking company, which I invested in. In 2004, he sold it to Symantec for $370 million. I pursued a consumer software product called MoveIt, which helped people seamlessly switch to a new computer.

Emmett Shear's Twitch

Before finding massive success with Twitch, Emmett Shear built and abandoned multiple products:

- Kiko—a JavaScript calendar that anticipated Google Calendar
- Youlookfamiliar.com—an ancestry.com-like family network (killed after three days)

> - Sounds App—a SoundCloud precursor
> - Justin.tv—the first "working" product, which led to Twitch
>
> For the first 18 months, the team killed any idea that didn't show immediate success after a week. As Emmett says, "It wasn't a strategy, but because we had ADHD, we were impatient and kind of impulsive. It was boring and it wasn't working so we moved on to the next thing we were excited about. . . . If you have to choose between 'most' and 'best' *shots on goal*, most is more important than best."
>
> For a product maker, ADHD is a feature, not a bug—but only when anchored to a singular vision. Rapid iteration helps you find what works, but discipline carries you through product-market fit. Be twitchy and give yourself permission to change direction as often as needed in service of that vision.

Finding True Signal

True signal is about tuning into that exact frequency that speaks to all of us. You feel it inside when a product nails it. Think back to the first time you tried the iPhone. It was exactly right, and it was magical. You didn't think, "Oh, this is close," or "It gets to a lot of what I want." Steve Jobs didn't think that, either. He knew the iPhone struck what was exactly right. That's why he didn't soft launch an MVP to see what we thought and then iterate.

When you and your team have this kind of conviction, you move at 10 times the speed, with 10 times the passion. After we launched *FarmVille*, we knew we had found true signal. It had 171,000 organic installs its first full day and a million its first week. The pace your team will move at is unstoppable. You're not asking anyone to work all night—you're making them go home to get sleep.

Sometimes, you may find the clearest validation of true signal when someone else executes perfectly on your instinct. The strongest signal I'd ever seen came in early 2004 when Sean Parker brought Mark Zuckerberg to meet me at Tribe. Zuck had the swagger of someone who already knew he had won the game. He was wearing basketball shorts and Hawaiian flip-flops; he was maybe 19 years old but looked 16, and he propped his feet up on the conference room table. His business card said, "I'm CEO, Bitch."

When Sean pulled up "TheFacebook," it was in four or five schools. He explained that it took a week to get 20 percent of the students and another week to get the other 80 percent. Sixty percent logged on every day—a metric that has held true to this day. I've learned over the past few decades that this 60 percent Daily Active Users / *Monthly Active Users* (DAU/MAU) ratio is the clearest sign that a service is truly part of people's daily lives.

Facebook spilled over with trust. College kids were sharing cell phone numbers and access to music files on their own PCs. The level of trust made my jaw drop. I saw that they had nailed everything I'd been trying to do with Tribe: They had taken the Cocktail Party instinct and executed it perfectly. My only solace was that soon after this meeting, I was offered the chance to invest in Facebook's seed round, which turned out to be the single best investment of my life.

Staying Close to the Metal

Any good CEO maintains a direct connection with their product and users—what I call *staying close to the metal*. In most cases, we're talking about software, but the "metal" could also be whatever the atoms are that make your product.

My favorite example is how in love Steve Jobs was with the fidelity of the glass on his iPhone 4. At Apple's 2010 Worldwide Developers Conference, he was walking me down the hallway with his arm hooked around my neck, berating me; he said FarmVille was failing to take advantage of the new crystal display glass because our game was built in Flash. He literally called his team, handed me the phone, and told us to figure it out and fix it that minute.

Staying close to the metal means constantly returning to your core instincts and never getting attached to the idea variants you're building. This is true whether you're in a consumer or B2B business—and my path to build Support.com is a case in point. As the incredibly roundabout story shows, our winning instinct was enough to raise venture capital and even get early customers, but it was only because I managed to stay close to the metal that I had the conviction to kill a gigantic losing idea that would have led to a zero outcome.

My first losing idea was MoveIt. But it led to a better idea based on the same It Just Works instinct I'd been chasing since FreeLoader. I called Scott Dale and Cadir Lee, the engineers I had worked with at SpaceWorks. I had tried to recruit them to FreeLoader, but they said it was too complex and would never work without a lot of bad hacks. They were right. But what they missed was the power of the consumer internet. Moving the open web an inch ahead turned out to be worth a lot. I explained MoveIt to them. "We like that problem," they said. "We'll get back to you."

A month later, Cadir and Scott sat me down and said, "Mark, MoveIt isn't that interesting, but you've identified the tip of a technology iceberg. It gets to a much more fundamental problem with Windows computing." They had an idea to build what they called DNA probes—software agents that would find the working state of any Windows computer and its apps, from Outlook email to Netscape browsers to your kid's *Carmen Sandiego* game. Cadir and Scott said this hadn't been done before, but it was theoretically possible—the perfect technology problem to pursue. The best software engineers want to build important, reusable software that can move the world forward.

It took Scott and Cadir a day to explain their idea, but then I saw the same huge potential, and we agreed to cofound a company. Though I started with a consumer idea, and everyone was crazed about the consumer internet, I realized that we should pursue an enterprise software business.

When we first tried to raise money, no one wanted to invest. Our technology was too technical for the VCs to understand. We got Accel to lead our Series A, though they remained a reluctant partner because they (and most other VCs) weren't excited about enterprise software. This was 1998. Anything with a ".com" was a potential big IPO. One time, they asked me to cut one of our partner presentations short to add more time for an Apple team building a graphical consumer user interface (UI) for Linux.

Our initial product addressed a problem that later became a category called electronic software distribution (ESD). While the product was not popular with VCs at the time, there was a clear market. The space even had a hot company called Marimba, started by Kim Polese, who was on the original Java team at Sun Microsystems. We generated early interest and sales, including a million-dollar order from Morgan Stanley's IT department. We used that deal to close our Series B.

At Support.com, staying close to the metal meant I was going on sales calls, even with zero sales experience. I found the average salesperson would say yes to anything to get a lead or close a deal. Our teams were selling ESD, because that's what the enterprise Chief Information Officers had budgeted. We fit in a category, and we could get orders. But that wasn't going to make us a profitable, scalable, successful company.

I realized that ESD was the Vietnam War of enterprise software. You could go in—and lots of companies did!—but you would never come out alive. The reason IT groups would buy was to get free consulting. These commitments would go on for years because the software was so custom and complex in these large corporations. These deals were actually fool's gold, requiring years of custom engineering that would drain resources and ultimately fail. (Spoiler alert: Every company in this space eventually went bankrupt or was acqui-hired.)

I had one of the loneliest, most painful wake-up moments a founder ever faces. I had led everyone—cofounders, investors, team members—up a losing hill. *Do I suck it up and hope it works out, or stop time and force us all to face facts with brutal intellectual honesty?* I chose the latter option, as I have throughout my career. That choice is part of how I've gained a reputation for being "controversial," as have many other founders now revered for their contrarian stances (after huge successes). Everyone prefers the "blue pill" life in *The Matrix*. Investors like confident CEOs who tune for harmony over the absolute best path. Sometimes perfectly aligned execution of a B+ idea can lead to a successful company—just not a great one.

Staying close to the metal meant also being close to my engineers. I constantly asked Scott and Cadir what was hard and what was easy. (This is a practice I continue to this day, despite endless teasing, because I can't write a line of code.) The sales process early on required a lot of "entrepreneuring." We needed to match customer pain/value with a repeatable application of our technology.

By going deep with Scott and Cadir, we developed a new insight. Our product could crush it in a new category of software we made up: help desk automation.

I joked that there was dust on the doorknob of the help desk manager in every company. But everyone had had incidents when they needed help desk support and were burned. No executive liked the helpless feeling of waiting on an IT person to regain access to email. And the cost of addressing what were called "break fix" issues was huge, from lost productivity to staffing. Our technology of DNA probes could deliver a magical experience in diagnosing what had changed in the working state of your application and either reverting to the former working state or finding the problem.

It wasn't an easy pitch to shift our entire company from ESD to help desk software, however. Cadir and Scott said, "Mark, you're totally right. We can build a great product and business plan, and there's revenues to be had. But let's do that with the next company. What you're trying to do is an unnatural act."

"We have to be intellectually honest with ourselves, with our investors, our customers," I said. "It's never too late to build the right product."

We had two days of brutal meetings where literally nobody in the room agreed with me. But it was my job as CEO and founder to be right, not to be liked. I knew I was right because I was playing every position in the company. There's a Russell Crowe movie called *Master and Commander* that defines what I thought great leadership was about. Crowe, as the ship's captain, goes down to the guys in the galley who are rowing, and he says something like, "If you row harder, it's to freedom. And if you don't, we're all going to be killed or enslaved." And then he grabbed an oar! That is the style of leadership I aspire to. People follow a leader who is in the trenches. That's different from micromanaging; it's being a *full-stack leader.*

My team followed me because they knew that I was in the details with them. The worst thing for a team is to have the feeling that the higher-ups are making disconnected decisions. Companies build a cynical culture when their leaders are not in touch with the data and the details, and politics are driving the decisions. That's why we have to all stay close to the metal all the time.

We spent another six painful months building a new experience to deliver what we called "self-healing software." We quickly learned that while this new core tech delivered magic and a clear return on investment, we needed to add more features, such as chat and ticketing, to sell a lot of software. (You need to deliver both measurable value and high user engagement to succeed at enterprise software.) This hard move saved us years of pain and eventual failure—in contrast to Marimba and all the other companies that went after ESD.

Keeping Your Emotional Center: Brian Grazer

As a product maker, you can't lose touch with your creation. Brian Grazer is the creator of some of the greatest movies and TV shows (including my family's favorite, *Arrested Development*), and he was also the producer of *Friday Night Lights*. *Friday Night Lights*, which reflected his experience trying out for the high school football team in a small town, going through Hell Week, and then getting cut as a wide receiver. He wanted to capture the emotion he felt in those 10 seconds when he went from "feeling like a human being to feeling like he wasn't a human being." He told me the film version was "an emotional connection" for him—so much so that it took him nine years to find the right director to make it the way he imagined it. But at some point, he stopped being directly involved with the television show and outsourced control to the showrunners to keep it going for more seasons. And even he says that the show lost that "emotional center." Viewers also lost the connection they once felt. Now, though, the series is being redone—and he and the original director and writers are involved again.

The Cocktail Party: The Ultimate Instinct Vein

Not long after starting FreeLoader and Support.com based on my IJW instinct, I formed a second instinct vein I still call the Cocktail Party. Reid calls this "the people web," and it's at the core of some of the biggest internet companies and industries, from Meta to TikTok.

We are all searching, online and off-line, for that perfect Cocktail Party—the gathering that combines curated people with the right context, purpose with fun, and trust with community. It's where interesting people, old friends and new, come together effortlessly. It's where we find our best "leads" for everything from dates to jobs, from travel ideas to culture.

This instinct vein has proved to be foundational to social networking and social media. But like all powerful instincts, it can lead to both winning and losing ideas. The challenge is separating the core human truth from any particular implementation.

This Cocktail Party instinct led me to found Tribe.net, which merged three social products in which I had observed true signal: Napster (peer-to-peer sharing), Craigslist (free local listings), and Friendster (social networking). Although only Craigslist was ultimately successful over the long term, each had nailed a truly compelling consumer experience.

The Matrix Moment: Napster

My first glimpse of the social web's potential came through Napster. Sean Parker, who had interned at FreeLoader at age 16, emailed me in 1997 saying that he and his cofounder Shawn Fanning needed $100K to buy more servers for their new peer-to-peer music sharing service, which was blowing up. (When you get that email, you just send the check!)

The first time I logged in to Napster changed my view of what the internet could be. It spoke to what I had always hoped the network could become but had never managed to articulate. The interface showed 4.5 million PCs connected to one another, each representing a real person, with no central service or corporate overlord. It felt somehow rowdy, even though there were no sounds or chats.

I typed "Madonna Like a Virgin" (embarrassing) and saw 17,000 files available across 10,000 PCs. This was mind-blowing. It was my *Matrix* moment. I felt I was looking through the network directly at all the other people around the world.

Up to that point, the internet had been a quiet, static library experience. We just followed links to pages, mostly served from big corporate-owned databases. Napster felt like the people's revolution.

Easter Egg

Build Another Napster

Twenty-five years later, we still don't have Napster. Ask a Grateful Dead fan whether they'd like a service that lets them see all versions of "Scarlet

Begonias" in one place. Ten out of ten would say yes. It would be the Cocktail Party for Deadheads, Phish fans, Swifties, and more.

It's become normal for us to record every live show, but what happens to our recordings? They go nowhere. Maybe you do this on the blockchain, and the music industry doesn't know whom to sue. This seems like a big opportunity that's still out there.

Communities want to share their passion for music directly with one another.

It's All About Lead Generation: Craigslist

I loved what Craig Newmark built with Craigslist. It proved that people could self-organize and replace massive corporations—in this case the newspaper oligopolies—with a simple, free HTML bulletin board. Beyond that, Craigslist proved that we could all trust and transact directly with one another. We didn't need intermediaries and professional agents.

Craig is the most humble and wise consumer product maker I've met. Craigslist, which has remained unchanged since it was launched on the web in 1996, is still the global leader in person-to-person classifieds, generating billions. Yet he has managed to keep it a small private company, where he continues to maintain his role as chief customer service representative.

Craig was my neighbor in Cole Valley, though he was actually better friends with my dog Zinga than with me!

The First Cocktail Party: Friendster

Back in late 2002, when San Francisco felt like a ghost town in the depths of the internet nuclear winter, I was getting together with people like Reid Hoffman to brainstorm about the future. We saw Amazon announce monster fourth-quarter numbers with 25 percent growth in consumer purchasing. While most people had given up on the consumer internet, we saw something different emerging: what Reid called Web 2.0.

The core idea was that all data on the internet should be free and move freely through application programming interfaces (APIs). What if the web could be made up of people, not pages? Larry Page and Sergey Brin had built Google on page rank relevancy. What if the future web was really about people?

This insight led Reid and me to find and invest the first money in Friendster, at a $500K pre-money valuation, which seemed crazy at the time. I made Sunil put up half my check.

Two early moments told me Friendster would change our world. I was sitting at a blackjack table at the Hard Rock in Vegas and overheard a college student from Ohio ask her friend whether she could "Friendster" her. My mind was blown. This little software app on a few random servers in San Francisco had made it to the middle of America and gained this much heat. The second moment was more personal. After Friendster had infiltrated my whole San Francisco community, I met and dated a woman through friend-of-friend links. This was mind-shifting. I had tried Match.com and found the dating "leads" so random that they became a painful waste of time.

The vision we'd been developing in those Web 2.0 brainstorms was starting to materialize. Friendster became the first social network, growing exponentially in months. A few months later we launched our own ideas: Reid created LinkedIn, and I built Tribe.net.

Easter Egg

Online Dating: Always Broken Opportunity

In 2017, after my divorce, a friend turned me on to Raya, which is a human-curated dating app. I had always thought online dating was a 1 out of 10 experience, generating such bad leads that it wasn't even worth meeting in person. Raya was actually a 3 out of 10; it sucked three times less, which meant you could meet people, not feel that meeting them was a total waste of time, and even sometimes get a good enough lead to go on another date. I cold-emailed info@rayaapp.com and, luckily, my friend Jared Morgenstern, who had become their Chief Operating Officer, saw my email. Soon afterward, I became a primary backer of what went on to become a very large business with amazing margins, since it relies on virality instead of paid customer acquisition. And yet this opportunity remains available to anyone interested in innovating. I recently invested in The Relationship Company, the new venture of a former Stanford student, Henry Weng. Henry's app, Date Drop, which leverages AI to offer college students a better match, has been used by 60 percent of singles at Stanford and is quickly gaining traction at MIT, Princeton, and Penn. I also love that he is taking advantage of the same organic trust network of email addresses ending in ".edu" as Zuck did in founding Facebook. By automatically connecting those whose email addresses are part of the same ".edu," he is offering an instant, large, trusted network.

Experiencing the Cocktail Party: Burning Man

Burning Man showed me the power of the Cocktail Party in real life (IRL). My 19 Burns over the past 21 years deepened my instinct that we all want to be seen by our communities as creative and useful. Burning Man, like Halloween, is a space that allows adults to play and express themselves in any creative way. Like joining a friend's cocktail party, it's about offering fun, creativity, and anything useful to the community. That was also the idea behind Tribe.net.

B+ Is the Enemy of A

Tribe seemed to be an instant success as it quickly became the online home for Burners and many online communities. By the end of 2003, we had a much bigger audience than LinkedIn. The promise seemed to deliver as people found roommates and amazing couches to crash on in faraway countries.

But I also received close anecdotal signs that this wasn't working. At the time, I had a preppy girlfriend from the Marina who was freaked out by people posting naked pictures and sending her unsolicited messages. I stubbornly (heroically) marched onward, like so many other founders who place their personal commitment above market feedback.

We were a *sinking speedboat*. We were getting lots of virality: An email from someone that said "Do you want to be my friend on Tribe?" had an 80 percent open rate and a 50 percent accept rate. The novelty worked. But while virality was there, the retention and engagement with anyone outside of some narrow communities was terrible. We had to choose between driving faster (more installs), bailing out (trying to keep people more engaged), or making the hole in the bottom smaller (improving retention). The rule I didn't know yet: Longer retention always wins.

I joke to this day that I managed to fail at a time when everything worked. Imagine launching a service at a time when people would actually open emails inviting them to be friends on a new online service and half of the time would actually accept the invitations! I started one of the first three social networks in 2003, the same time as LinkedIn, six months after Friendster, and a year before Facebook. If I hadn't stuck to one idea but had instead tried an alternative second, third, or fourth idea and gone with whichever one generated the most heat, I might have built something you've actually heard of instead of Tribe.

The fundamental problem was that I had three winning instincts trapped inside one losing idea. The fatal flaw in the concept was trust; I just got it wrong. Trust is so fundamental to the Cocktail Party; you can't get it wrong. You do not get to pass go if you do not have trust in your system.

I made another classic mistake by chasing false positive signals. When *The Washington Post* and the large media company Knight Ridder wanted to invest, I thought, "These big smart corporations think we're the next big thing!" I believed the newspapers had no alternative but to promote us because their classifieds business was dying. I didn't realize they had another option, which was just to die.

We had a very corporate board that included two directors from media companies (*The Washington Post* and Knight Ridder) and one VC. In 2005, they said they would not fund the company any further unless we hired a "real CEO." The guy we got was a born-again Christian who, oddly, spent all his time trying to police the adult photos on the website. By 2006, Tribe was out of funding. We sold the technology and software to Cisco for enough money to pay back our venture capital debt.

After Tribe, I went into a multiyear Abyss. I pursued an idea called TagSense—a different kind of B+ trap. The concept seemed solid: a publisher-driven ad network where blogs and websites could customize their AdSense experience through keywords. Peter Thiel liked it enough to commit $5 million, and publishers wanted to be first pilots.

When you're trying to get out of the Abyss, a B+ idea can feel like an A. But attracting smart investors doesn't equate to true signal. Who knows? Maybe Peter was thinking I'd eventually turn TagSense into something better. But the idea itself wasn't that good. It was "meh." I pulled the plug before we even took the money or launched anything real.

Whenever I ask founders, "How ambitious are you on a scale of 1 to 10?," everyone rates themselves a 10. But very few are ready to make the trade-offs required to create a truly winning product. Perhaps the hardest challenge for all of us as product makers is to have the intellectual honesty to say, "This idea is a B+, and it's not going to be a hit."

Being truly ambitious and committed to winning means:

1. Maintaining ruthless objectivity about your ideas.
2. Being willing to kill them often.
3. Losing any emotional attachment to specific implementations.
4. Testing multiple variants quickly rather than going all in on one.

When you're trying to build products, good is worse than bad. B+ ideas are good enough to get funding, get customers, attract talent, and keep hope alive. But they're never good enough to be great. You can avoid good and get to great by asking questions: *Would your smartest friends invest their own money? Are you pursuing this idea out of desperation?*

Developing Taste

The greatest product makers have great taste. I don't trust a chef who doesn't have a point of view on their own cooking, and similarly, I don't trust a product maker who doesn't have a strong perspective on their product.

This isn't about having universally appealing taste; it's about developing a keen sense of quality and maintaining high standards. Product makers who don't use and deeply understand their own products are like chefs who don't eat their own cooking.

I've spent my whole life trying to make the perfect Bolognese—not a recipe I follow, but a sauce I make up every time, iterating and adjusting. I know when I love the sauce, other people will love it, too. I know I've hit it when my kids ask for thirds and eat it cold for breakfast the next day.

This isn't about sauce, though if you're interested, I've included my recipe below. This is about developing the ability to recognize when something truly works. As a product maker, you are a tastemaker and taste tuner. It isn't a matter of being arrogant about your taste. It's about being a student of your own reactions. When you lose interest in something you thought would be huge, don't ignore that signal. When you can't stop using a product despite its flaws, pay attention. Your taste is your compass, if you've calibrated it properly.

Great taste comes from the following:

1. Deep immersion in your product category.
2. Constant experimentation and refinement.
3. Regular exposure to user feedback.
4. The courage to maintain high standards even when it's inconvenient.

Mark's Bolognese Sauce

It's not quite right to call this a recipe because I made it up every time. But here's how I make the perfect Bolognese.

Ingredients

2 pounds lean ground beef (about 1/4 pound for each person you are making the sauce for). I always try different kinds of ground beef—tenderloin, New York strip—or I play with the fat content.

Garlic cloves to taste

1–2 onions

Olive oil

2 packs of sliced cremini mushrooms

Butter

3 small cans of tomato paste. Look on the label to find ones that are low in sugar, because sugar is the enemy of your sauce. You cannot use anything that calls itself tomato sauce. That will have 30 grams of sugar per serving, while a small tomato paste will have 6–8 grams of sugar per serving.

Italian table wine. Whatever you want—a wine that you want to sip while you're making the sauce. I don't like pinot noir in my Bolognese.

Italian tomatoes that come in bunches

Seasoning: garlic powder, sea salt, bay leaves, oregano, Parmesan cheese

Important note: Do not add carrots or orange juice or anything sweet. Sugar or anything sweet kills a Bolognese.

Instructions

1. Sauté lean ground beef until rare to medium rare. Dice and lightly grill garlic cloves. Chop 1–2 onions. Lightly sauté in olive oil on low heat. Don't cook the onions too long or they will get caramelized and sweet. Lightly sauté the mushrooms in butter.

2. In a large pot, combine cooked ingredients. Add 3 small cans of tomato paste and water. Pour in wine. Put in tons of garlic powder. Add sea salt and bay leaves to taste. Keep dicing tomatoes and tossing them in as needed. Simmer with lid on for at least 2 hours with a fair amount of water.

3. Spend the next few hours doctoring the sauce and seasoning it to taste. If it's flat, add salt, oregano, and garlic as needed.

4. Chop another onion, lightly grill it, and add it to the sauce. If the sauce gets too briny or too overpowered by the meat and saltiness, add tomato zestiness. First, try more fresh diced tomatoes. If that doesn't do it, add a can of diced tomatoes. Possibly add a little more tomato paste, but the tomato paste can make the sauce too thick—like ketchup.

5. It's got to be saucy, so it sticks to the pasta everywhere and you don't have lumps of meat and pools of liquid.

6. Thirty minutes before serving, add grated Parmesan cheese to the sauce.

Distribution Trumps Instinct

During my post-Tribe Abyss, I found myself expressing a frustration that had been building since FreeLoader. I wished I had a captive audience that could allow me to test ideas quickly. I didn't want to go build another Tribe; I couldn't spend a year building plumbing just to test one idea.

The real problem wasn't just distribution; it was the fundamental cost of testing ideas. Every time we wanted to try something new, we had to build infrastructure (plumbing), raise venture money, and spend months building teams—all before we could test whether we had true signal.

My friend Matt Ocko pushed me to think bigger. What if I bought CNET, which had something like 30 million MAU in the US alone? Then I'd have a captive audience to test tons of ideas. Matt connected me with someone who helped me talk with private equity firms. I moved to New York and began trying to raise money to take CNET private.

Meanwhile, I had a small team running what remained of Tribe. I kept a skeleton crew working on small, scrappy ideas. This experience of running lean and trying things quickly was teaching me something, even if I didn't fully understand it yet.

The CNET plan was another B+ idea, good enough to pursue but not quite right. The instinct behind it was gold, however. I was starting to see that the real innovation needed wasn't just in what we built, but in how we tested and validated ideas.

In late 2006, I went to Dialog, a conference Peter Thiel and Auren Hoffman had created to have the kind of conversations you couldn't have at normal tech events. I ran into Matt Cohler, who was running corporate development at Facebook. He told me they were opening APIs for their platform: "We're looking for anyone we can find to build apps for it. You should take a look."

My mind started racing. I had invested in Facebook, but more than that, I had this deep instinct about platforms and APIs from the Tribe days. We had seen it coming. There was going to be an arms race as social networks tried to build every possible application themselves. At Tribe, we had even built a plug-in architecture because we knew people would need to add things like video.

This was the promise of Web 2.0 that Reid and I had brainstormed about back in 2002 and 2003. We shared a vision that data would be free, everything would have APIs, and users would just integrate whatever they wanted. We could see it starting with the photo-sharing service Flickr.

Instead of pursuing CNET and burning time raising money to buy a public company, dealing with bankers and boards just to get a testing ground for ideas, here was Facebook about to hand me—to hand everyone—a captive audience of 70 million users. There was no infrastructure to build, no plumbing required; it was just pure experimentation at speed.

Every founder faces this same challenge around distribution, so we all need to innovate on cheaper ways to prove signal and potential for our products.

TL;DR

We can change our odds of success by focusing on our instincts, which are right 95 percent of the time, but killing our ideas, which are right 25 percent of the time (at best). An instinct is a deep human insight—such as knowing phones and computers should merge, or that getting a taxi should be easier—while the idea is the specific instantiation, like the iPhone or Uber. Most entrepreneurs fall in love with their first idea variant, but that's rarely the winner. Instead, pursue your instinct veins—those fundamental insights about human needs that can spawn multiple products. The deepest instincts often form during an Abyss, one of those painful periods when you're forced to slow down and really observe. Don't rush past those moments. They help you develop the conviction that will sustain you through the iteration required to find the winning idea.

3

The MVP Trap

Over the past 20 years, we've all been sold this concept of Minimum Viable Product (MVP) as THE way the best teams build and launch products. The problem with this concept is that we no longer have the time to deliver on "viable." My friend Eric Ries, who turned MVP into a religion with his book *The Lean Startup*, defines "viable" as real users can utilize your product for real-world behaviors.

Eric's objective with *The Lean Startup* was to find ways for teams to get to market faster to accelerate their learnings. But in practice, viable has become an excuse for teams to invest too much time and money in their product before ever launching (and learning).

After years of listening to teams tell me their MVP isn't good enough to launch, I've come to the conclusion that most teams overinvest in the hope that viable translates to successful. The reality is that the chance that an MVP—the first thing you put in the market—is going to succeed is close to zero.

Lately, I've been encouraged to hear the scrappiest founders say if you're not embarrassed, you're not launching early enough. But that remains the exception, not the rule. What we need is a new mindset and framework—a smaller, less ambitious atomic unit.

Let's call it **Minimum Idea State** (MIS). MIS isn't a whole product; it's just enough of an idea to put in front of your friends and target audience so that you can get useful directional feedback. These can be any instantiation of your idea, from a text link to a PowerPoint to a vibe coded piece of demoware that fails after 10 clicks.

With Zynga, I didn't have the time or the ***engineering days*** to be wrong. I had to prove strong user interest before building anything. We displayed five-word links to see what percent of our users clicked, even though they went to 404 Page Not Found. Once we proved the idea had heat, we quickly built the first few levels of a game to generate initial engagement and retention.

In 2008, at Zynga, we were trying to grow Poker on Myspace, but it lacked any viral channels. Still, we had managed to build Mafia Wars to one million DAU. I was drinking beers with some engineers one Friday night and convinced one of them to help me change the text and link on the Mafia Wars home page. We tested "rob them at the poker tables," which we made publicly accessible for one minute. We found that 22.7 percent of Mafia players clicked the link. We turned it on again, for another minute, and saw 22.7 percent click. That gave me and the Poker team enough confidence to take 24 hours (which was a long time then) to build Mafia Wars–themed poker tables and link them to the home page of that game. When we turned that on for everyone, 22.7 percent of players clicked, which in one week led to one million installs of Zynga Poker.

Most successful consumer internet companies have some similar story about their own version of getting to Minimum Idea State—testing their idea before wasting cycles building it right. Drew Houston proved his concept for Dropbox by creating and testing a two-minute demo video showing an imaginary product, which generated thousands of sign-ups. The DoorDash founders put up a web page advertising delivery from local restaurants, took orders via Google Forms, and then did the deliveries themselves. And finally, for Zappos, Nick Swinmurn proved the concept by posting pictures of shoes from his mall on a website and then buying them at full retail and mailing them himself to customers.

The best example of this today is Rollic, one of the leading hypercasual gaming companies, which Zynga later acquired. Rollic moves so fast that they're able to build what they call "TikTokable" games. They pair popular game mechanics with top trending memes. When we acquired Rollic, they had a game called High Heels, which was the number one free game globally in the App Store for several weeks.

Today with AI, we have the ability to go from a prompt to a prototype that conveys the end-user experience. Both engineers and noncoders can vibe code actual working versions of most product ideas in minutes or hours. These experiences aren't an MVP, they're Minimum Idea States—they're closer to working prototypes or demonstrations, rather than shippable products.

Why am I making such a big deal out of this? It's important that we break our product down into these much smaller atomic units that we can learn from quickly. Ultimately, we want to avoid wasting weeks and resources on a product or features that nobody ever wanted.

Minimum Idea State Through Ripomatics

One of the easiest ways to get to a Minimum Idea State is to build a *ripomatic*. TV and film studios originated this term to describe movie trailers, which allowed them to convey the tone, style, and mood of a new project before production started.

The idea of a ripomatic is to tell the story of a user's experience without touching code. Often, consumer products are mashups of other successful features. A ripomatic pulls these disparate experiences together and shows the step-by-step user experience—how they use and interact with your product. User interface is less important, but the user experience should be clear and obvious, using realistic copy and mock content when possible.

At Zynga, we asked teams to get their idea into a PowerPoint, with screenshots and video captures that they could put in front of users. That was often enough to convey the concept to users and get useful feedback.

I asked Johnny Li, who ran product at Erth.AI from 2022 to 2025, how he defines and uses ripomatics. "To me, a ripomatic is like putting together a scrapbook. Instead of creating wireframes from scratch, it accelerates the concepting process as you pull in screenshots and recordings from existing products. It helps to mash them up together with annotations and overlays to demonstrate the new product concept. This also ties into Proven Better New, since you're bringing in a lot of Proven from existing products." (You'll find out more about how Proven Better New helps you take the most best shots on goal in the next chapter.)

Today, with AI and vibe coding (in which you can literally talk to a large language model and have it code your ideas), we are close to being able to build fully interactive ripomatics that verge on whole product prototypes. This massively improves on my old PowerPoint and static screenshot approach! There may be a risk that these ripomatics are so good that they confuse end users, who expect to have fully functioning software, but it will be amazing to see how far product makers push this.

We should all be pushing our boundaries with vibe coding. Reid keeps reminding me that the winners will bet that vibe coding exceeds expectations, just like the massive shift in use cases we saw from GPT-3 to GPT-4. The point is that GPT-3 was mostly useless, but it pointed to a huge vision of what could come soon. In fact, it led Emmett Shear to tell me that a great target for launchable user experiences is delivering something magical for a small number of users. If 20 people really love your product, that's better than one million thinking it's just okay. Similarly, we could bet that vibe coded applications will be good enough to launch real products because they point to a huge vision, even if they won't actually deliver for a long time.

Case Study: MIS Saves a Sinking Ship

The power and speed of building for MIS over MVP was proven with Words With Friends in 2015.

The game was in decline, and the team was looking to reverse the trend. They had this assumption that users found the game too slow. So they decided to build a "Fast Play" game—a smaller board that generated faster gameplay and lived alongside the classic game. The core problem, and the reason the effort was doomed from the start, was that a team of 20-something men were working on a game whose primary users were middle-aged women.

The new game took six months and hundreds of engineering days to build. When it finally launched, only 5 percent of users ever tried it, and none returned to it. It was a complete flop. Our female players hated it. They didn't want a faster game; they loved the Zen of a turn-based game that they could come back to any time.

When I found out about this in a roadmap meeting, I asked the team, "Why didn't you just test this with a dialog box before you built the game?" This test would have taken one engineer and one PM around an hour, versus thousands of engineering hours.

We had no time left to be wrong, so we moved to this MIS approach, and it ultimately led to a winning idea: Weekly Challenges. We tested it quickly by putting a minimal version on the web—a progress bar that filled as players earned more points. When this showed positive retention signals, we built it into a full feature and continued to iterate.

Weekly Challenges helped us save Words With Friends and, with it, Zynga. We went from $80 million and rapid decline in 2015 to $200 million and growing in 2016. The key to this amazing turnaround was our ability to build huge conviction with one small test.

Case Study: Sitcom Takes Off from MIS

MIS is not just for games or software. Comedy writer and producer Dave Goetsch told me how he created just four scenes for his new sitcom instead of an entire pilot to prove signal and get the right partner on board.

In 2018, when *The Big Bang Theory* announced that the twelfth season would be the show's last, Dave began to think about what kind of project he wanted to work on next. He had an idea for a new series about a US Marine who had served in Afghanistan and returned home to Ohio. When the story begins, the Marine's Afghan interpreter, someone he had developed a deep bond with during his deployment, immigrates to the US to join him. Dave's writing partner, Maria Ferrari, knew that their best chance of getting the show on the air was if they could bring the "king of sitcoms," Chuck Lorre, on board.

A traditional approach would have been to produce and write the entire script, but having worked with Chuck for years, the writers decided their best shot was to present something concise but powerful: a few scenes that conveyed the essence of the show. They wrote just four scenes to capture the heart of the story and the chemistry between the main characters. When they handed the pages to Chuck, they told him that these were "moments" from the world they wanted to create. Some might end up in the pilot, others might not; but together they expressed what the show was about. This was a perfect ripomatic.

It was enough to convince Chuck to do the show with them. Together, they went on to write a pilot for *The United States of Al*, sell the series to CBS, and produce two seasons, totaling 35 episodes, a longer run than most comedies manage today.

Dave says that if they had written an entire script, it might have given Chuck more reasons to say no. By focusing on the essence of what they wanted to create, they made it easier for him to say yes, and they were able to deliver those scenes—and move the project along—months earlier.

Maximum Launch Product

The real goal of the MIS is to get to our Maximum Launch Product. We want to find true signal with real users, so that we have the conviction to overinvest in the product.

In later years at Zynga, we were over-resourced and wasted millions on new games. Every team would tell me their game wasn't ready to launch yet. They hadn't "found the fun." They'd delay, go massively over budget, and always think they needed just a few more weeks. But here's what they didn't understand: The level of creative productivity you have once you're in the market, interacting with real users, is 100 times greater. You can never test early enough or raw enough. I still struggle with this lesson today.

Until we've tested, every idea is possible. And that's the problem. Any idea is just as good as any other. But after we've tested, we see that some ideas are winners and most ideas are losers. Once we get to the winning ideas, then we consolidate around those. We move much faster and stop wasting time on features that nobody wants.

Brian Chesky, the CEO and cofounder of Airbnb, is a master of testing by hand. He and his team can move faster by testing ideas anecdotally on small numbers of people and building signal off that. He's gotten so good at this that he is able to build the conviction to never launch MVPs and to launch only Maximum Launch Products. I do something similar by constantly running ideas by people in my target audience when I find them out in the wild. I've always thought it's similar to how a stand-up comic develops their working routine. Seinfeld said it took him a year on the road to develop 45 good minutes.

Kill Hope Before Hope Kills You

In Silicon Valley, many people say, "Hope is not a strategy." Yet, when it comes to launching products, too many founders produce hope shots instead of conviction shots. By investing too many resources and too much hope in their MVP, they miss the opportunity to fail fast and make micro or macro adjustments along the way, so that by the time they do launch they have built real evidence that their product will resonate.

Hope is a belief that's not supported by any data. As product founders, we feel the need to express confidence around the product we're building, to inspire our teams and investors. Too often, we get over our skis and fall in love with the possibility of what the product can become long before we've built the hard-won proof of and conviction in our ideas. This is why it's so important to find these early proof points that build real momentum in the right direction.

There are many ways great product makers go about proving to themselves that their product is right. Brian Chesky does this "by hand," meaning he will literally create the experience for himself before bothering with software. Most famously, he and cofounder Joe Gebbia hosted people in their own apartment to test the idea for Airbnb. I have done this with ripomatics. You can even do this through telling a good story and observing people's responses.

The best product makers determine they have a winning product long before they release it to the world. They leverage instincts but prove out their ideas so that they reach deep conviction, then launch in the biggest possible way.

Case Study: Did Hope Kill Quibi?

Quibi is one of the biggest recent examples of a talented, highly proven entrepreneur betting the farm on one industry-changing idea rather than maximizing shots on goal. Jeffrey Katzenberg invented the modern animated film genre, creating an incredible series of hits at Disney and later at his own company, DreamWorks. We were lucky that Jeffrey was an early believer in Zynga and joined our board, where he pushed us all to think bigger. While Zynga was proud of its leadership in big data, Jeffrey was ahead of us in using audience metrics to predict a movie's entire gross revenue based on opening weekend numbers.

With Quibi, Jeffrey made a massive bet ($1.75 billion) on a brilliant instinct that the next generation would prefer to consume video entertainment in snackable bites on their phones. This was already happening with user-generated content (UGC) videos on TikTok, YouTube, and Instagram.

I remember talking to Jeffrey early on and offering to help on product management and strategies to test ideas. I felt some guilt that I couldn't convince a friend he was skipping valuable steps.

The problem was that Quibi produced a Maximum Launch Product without proving signal first. The company shot it all on one approach—being the Netflix for phones, with celebrity talent and traditional Hollywood production values—rather than testing cheaper, episodic formats that might have revealed what actually resonated. Quibi committed too hard, too expensive, too early.

Quibi pursued one bold idea, got it wrong, and shut down the service after just six months. Jeffrey's explanation was, first, that the pandemic erased the company's intended use case of serving commuters and, second, that they didn't get product market fit right.

The painful irony to me looking back today is that Jeffrey's instinct was exactly right. Within a couple of years, the explosion of one-minute serial video apps such as DramaBox and ReelShort proved his thesis. These apps now generate billions in revenue in China and the US and are growing rapidly. If the Quibi team had anchored on their instinct but were less attached to their specific idea of how it would manifest, they could have taken more shots on goal. They might have found true signal and discovered that users wanted cheaply made, addictive serialized content rather than high-production mini-shows.

Build Failure Machines

We need to get to winning ideas on top of our winning instincts. The fastest way is to build failure machines, by finding new ways to test and kill large numbers of losing ideas.

There's a paradox we all face as product founders: the trade-off between the need to fund a team and test ideas, and the pain of raising VC money ahead of product market fit. We need to express confidence that our idea is going to work, but we don't have any reason yet to believe it will. This paradox is what motivated me to try to buy CNET in 2006—to have a captive audience I could test rapidly against. And it's what made me so excited to build apps on Facebook rather than bothering with the multiple billions needed to buy an aging internet company.

In 2007, I was starting to move toward the concept of the Minimum Idea State, which became the foundation of Zynga. My idea was to build a social poker game cheaply and then to test it quickly. It was like speed dating ideas—finding concepts compelling to us in the moment and exploring whether they had potential, without the burden of building a whole product.

The only reason I broke through the MVP trap with Zynga was by building small apps cheaply on top of Facebook. If I had to raise venture capital and get to launch metrics, it wouldn't have been fast enough to get the idea off the ground.

Easter Egg

Hack Distribution

Too many founders naively buy into the premise that if you build the best product, you'll win. In actuality, many of the most successful consumer internet services were built on top of a distribution hack. YouTube's first distribution was on Myspace, where users posted embedded links.

Today, the most overlooked mature distribution channel is Snapchat, which with 470 million DAU (as of Q4 2025) is even bigger than Twitter/X. And, even with all the hype, I'm surprised how few companies are hacking generative engine optimization (the new search engine optimization) on OpenAI, Gemini, and other AI platforms. The point is that many winning ideas start with solving distribution.

Don't Even Start with an Idea!

Emmett Shear has an even bolder approach to replacing MVP: Don't even bother starting with any specific idea. His insight is to use data to generate your idea, instead of using your idea to generate data.

This approach goes against everything we've been told but is so powerful. The moment you show customers something and ask them to critique your work, you're already compromised. Your ego is tied to your product. You've mentally invested. You're trying to prove yourself right.

According to Emmett, "Talking to the customer cannot improve the ideas you've already had. You can only improve new ideas you have after that. And people are really bad at throwing away their old ideas." So how do you avoid getting stuck on your bad ideas? "Just don't have any ideas. Don't have the idea until after you've gathered some data."

We need to flip the script entirely. Assume we can't talk about our solution at all. Start by asking people what they did today, what problems they face, and what frustrates them about current solutions. Then be quiet and listen.

This approach drastically increases your ability to execute because it reduces the marginal cost of getting a shot on goal.

Most or Best Shots on Goal?

Emmett made a compelling case that we should focus on *most* shots on goal, because none of us knows which will be *best*. The core principle I agree with is that it's about speed, not perfection; it's about learning before committing.

The tension between speed and quality has always been at the heart of product making. This is what makes the MVP trap so alluring and hard to escape. We waste so much time building a viable product when we should be solely focused on building a learning product. I'm guilty of that today with Erth.AI.

TL;DR

The problem with the Minimum Viable Product approach is the word "viable"; it becomes an excuse to wait too long and invest too much. Instead, get to a Minimum Idea State, with just enough development to either kill the idea or, hopefully, find signal. The goal is to get to conviction. Test every idea as early and raw as possible.

4

Proven Better New

The process of defining, testing, and launching new products can feel chaotic. How do we decide where to innovate and how to build conviction? How do we increase the odds of a hit and decrease the time to get to "right"?

In building Zynga, I had to be right to survive—so I developed Proven Better New. It's a simple framework that I've spent my career trying to master. It became a religion inside Zynga and has also spread to many of my fellow product makers.

Defining Proven Better New

Proven Better New offers a systematic approach to generating winning products faster. The core value is in isolating truly new ideas and then testing them rapidly to get accurate customer feedback. When used well, this framework helps us build conviction much faster by getting to our Minimum Idea State with greater intellectual discipline. Because this framework protects us from our urge to make everything better, we avoid wasted cycles and false signals.

Proven Better New also helps us avoid the time to reinvent necessary features and eliminates the false signal generated when we inevitably get those features wrong. In other words, we need to get comfortable copying what already works (legally), so we can spend all our time and energy on the novel innovation that actually excites us and our users.

Proven Better New Framework

= **Proven.** A product that is already loved by your target audience— the exact features/mechanics (every pixel) that make up the experience.

> **Better.** What you can improve so that 10 out of 10 users would respond with *"Fuck yeah!"* Be careful—what you think is *Better* is actually *New*.

+ **New.** Your novel idea that's never been tried with this product and audience. This is the riskiest part, so you want to isolate that idea and test it.

Why doesn't everyone use Proven Better New? First, not everyone is aware of this tool yet! But there's also an inherent moral arbitrage in really going after the Proven. We all want to be respected by our peers, so our egos hold us back from copying—as does our lack of experience as product makers. Copying what works goes against every instinct that we have as product makers: We want to innovate! The idea of copying other people's work can feel distasteful, wrong, unfair, and not something we'd take pride in. In my experience, the more junior the product teams, the more they spend cycles trying hard not to copy or even look as if they were copying.

The greatest artists and greatest product makers have usually followed a path of being the best at copying what's already out there; they've stood on the shoulders of giants. Steve Jobs, the grandmaster of product making, famously said, "Good artists copy; great artists steal." (Ironically, he actually stole this quote from Picasso, who was a master at tracing and copying other artists' works before he created his own.) Only a master craftsman can spot excellence and be comfortable copying it. They recognize when something is already excellent and understand that changing anything will just make it worse—and they're confident in their ability to create something new and innovative that will speak for itself.

I always say that if you're truly ambitious, you need to burn your résumé. Stop worrying about getting respect from your peers and industry. If you view innovation through the eyes of your customers, it won't be impressive to the rest of your industry.

Steve Jobs also spoke to this when he returned to Apple in 1997 and addressed one of his first developer conference audiences about the new Mac. "This whole notion of being so proprietary in every facet of what we do has really hurt us. It might be 10 percent better, but usually, it ends up being about 50 percent worse."

If all we did as product makers was copy, people would have no reason to use our product instead of whatever we were copying. That's where our innovation zones of "Better" and "New" come in.

"Better" refers to something in the Proven product that you can do better in the eyes of the current audience, so that 10 out of 10 of those users will say, "Fuck yeah, that's better!"

"New" refers to the novel idea that is our real innovation zone. It's something so new that it will get people to try, talk about, and even love our product. At the same time, we have to acknowledge just how hard it is to get an actual new idea that the world is ready for and wants. One of my maxims inside Zynga, repeated to this day, was "all New fails." It's not to discourage innovation, but rather to encourage experimentation and curiosity, rather than presuming success. New is the shiny penny everyone wants to spend all their time on. We put all our hopes and dreams in imagining how our new idea can change the world. The disciplined product maker knows to isolate the new zone to derisk their product.

We need to come up with a process to do enough Proven that our product doesn't fail for the wrong reasons. We need to find Better if we can, because it will increase our odds of success. Then, we need to try lots and lots of isolated New ideas on top of our product until we get to something that does work. This process of isolated innovation will help us massively change our odds of success to come up with our winning product.

Zynga Poker: Pure Proven Better New

The Proven Better New framework came out of the pain and failure of Tribe.net. I was so determined not to repeat Tribe that I prosecuted Zynga with a totally different mindset. I had a deep instinct that games could become a daily habit for busy adults and, eventually, one of the largest industries on the planet. Adults needed permission to play, which came in the form of being social and therefore somewhat productive. Games had always been "empty calories"—never considered a good use of time. I knew social gaming could be massive, but I didn't know the idea variant and product that were going to prove that.

Ironically—and this is the way life sometimes works—while I was ready to pursue many variants dispassionately, the first one I tried worked. That was Zynga Poker.

The fastest way to understand Proven Better New is to see the breakdown for the original Zynga Poker game from 2007. We took the Proven table layout and game mechanics from online poker, made it Better by removing the download, and added a New feature, which was real people, often your friends, with their pictures at the table.

Real Money Gambling Poker

Zynga Poker

Zynga Poker image courtesy of Zynga, Inc.

The Forces That Drove Proven Better New

As product founders, our best ideas are often formed from an extreme need to be resourceful. When I started Zynga, I was funding the company myself with only $350K because I didn't want to go back to VCs hat in hand after my failed Tribe experience. It was crucial that I get to breakeven on my own. I had to get something out fast that would also have a chance of resonating with users.

With only one engineer, I was forced to be much less ambitious. I couldn't build much, so I was focused on a narrow innovation zone. And without knowing it, I was using Proven Better New before I even had a name for it.

With Tribe, we were exploring such gigantic new land; we reinvented everything we touched. We offered social *and* professional profiles, along with groups and classified listings, while Friendster focused only on social and LinkedIn only on professional networking. Tribe was wide and shallow, complex and confusing.

Zynga Poker was the antithesis of Tribe. The product was so unambitious and noninnovative that it was embarrassing. It was just a poker game that looked exactly like every other. But here's what made this a winning product. We made two simple changes that turned our game into the only live Cocktail Party on Facebook where people could play and interact with other real people, often their friends. Our one New feature was showing players' real identities with their Facebook profile pics instead of random creepy avatars. Our one Better feature was instant play with no download, because we built it in Flash for the web. No other games could afford to offer web/instant play, because they all offered real-money gambling, where security was crucial. We had an advantage because we were pursuing a free model. To call out why this was such a big deal, I've always found you lose 80 to 90 percent of your audience when you require a download. It ultimately didn't matter that some players found ways to hack our game. But it was a bit embarrassing and funny when in our first month a kid figured out how to turn our whole table into a dick pic. By isolating our New aspect to purely social elements, we could test our core hypothesis without reinventing everything else. When we saw 25 percent of people were joining their friends at tables, we knew we were onto something big.

Proven Better New Became Zynga's Growth Engine

The innovation and growth engine at Zynga was the invention of social game mechanics. It's worth double clicking on what I mean by a *mechanic* and why they're so powerful, not just in games, but across all consumer products. A game mechanic is an interaction that's so fun and addictive, players enjoy doing it over and over again—for example, in Words With Friends, placing letters that spell a high-scoring word. Addictive mechanics show up everywhere, not just in games—likes and follows are baked into every social media product. The key insight is that once a mechanic is proven, it becomes a reusable building block.

At Zynga, with social game mechanics, we built something no one had offered before: the integration of a player's social network into their game board. This was how we disrupted the game industry.

My innovation roadmap for the whole company was on a whiteboard I leaned against the wall in the YoVille office. We wrote down all the proven game mechanics we'd ever heard of: experience points (XP), achievements, quests, points, badges. None of us really knew that much about them other than as players. But we saw that every single one of these game mechanics worked, so we just kept trying the next one.

These mechanics weren't available yet in mass-market casual games, but they had been well proven over the 25 years in console and PC games. They all worked, meaning that the same core loops were addictive. (A core loop is the primary function of your product—in FarmVille, it was plow, plant, and harvest. In Instagram, it's posting a photo and generating likes.)

In addition to learning from proven game mechanics, we constantly benchmarked the rest of the social gaming ecosystem. If another game generated better growth or engagement, we would deconstruct what mechanics they had implemented. Then we would match their experience and innovate on top. Our mantra was to be "best to market," not "first to market."

For example, with FarmVille, we launched a game that was Proven, Better, not New compared with the leading farm game, FarmTown. But within weeks, we launched innovative new features such as "buildings that matter," which created a deeper game inside your buildings, and soon afterward "animals that move," which led to years of new gameplay.

Millionaire City, an innovative game from the team that eventually went on to build Supercell, included a novel mechanic: Players could hire their friends on their game board. You could hire your friends into city positions, and if they performed daily tasks, you all progressed faster. We called this social game mechanic the "Crew Mechanic," and it enabled our fast follow game, CityVille, to grow to 100 million MAU.

Case Study: Slack

In 2012, Stewart Butterfield realized he had to make a hard pivot. His company, which had been building a massive multiplayer game called Game Neverending, was low on money with no prospects of more funding. People at the company asked themselves, "What can we do?" and realized they had this one internal engineering tool they all used. That was Slack.

Slack was a perfectly implemented Proven Better New against HipChat, the team collaboration tool that was already out there. HipChat was a popular chat product used inside most large enterprises to manage internal comms, primarily between engineers. Here was an enterprise product everyone had relied on that wasn't fun. Because Slack was the product of a game company, they built a more cuddly, fun front end on top of what was functionally still HipChat. Slack's Proven Better New strategy was to offer the exact same product but make it better in that the interface felt more

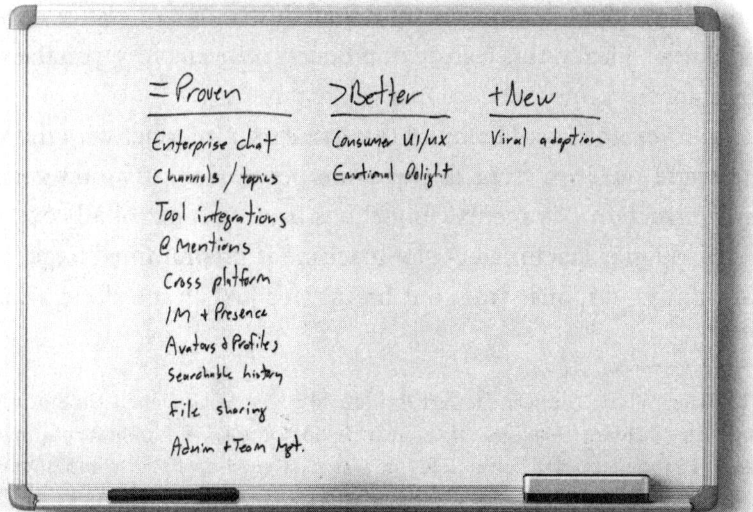

friendly. I don't even remember whether they had anything New initially. I do remember that the product greeted you with 25 different words for hello.

When the company launched the product as Slack, it spread virally: Eight thousand companies signed up within 24 hours.[*] Everyone loved it—but they couldn't explain why. Inside Zynga, our IT group was rolling their eyes: "There's nothing different. Why are you forcing us to get this?" But the big idea of Slack was making very small innovations that were very meaningful to the end users (not to the IT department). The product spread through work groups and surpassed HipChat in a year. Eventually Salesforce bought Slack for $27.7 billion, making it one of the largest acquisitions in the software industry at the time.

= Proven: Be a Heat-Seeking Missile!

In consumer markets, it's basically impossible to predict what will work. That's why we change our odds by starting with what already has heat—what is already *proven*. Think of Proven as your market entry strategy.

To be a master at Proven Better New, you need to get your PhD in *deconstructing* related Proven products that have heat. I've always believed that you don't have the right to build anything new until you've mastered what has already worked. Similar to game mechanics, we can deconstruct every feature of a product that makes up the totality of its experience down to the pixel level. Ideally, this feature or product has been proven on the same platform for the same audience that you're pursuing.

"Proven" means I could rebuild that same user interface with the same features and put it in front of the same people, and it would generate the same reaction. My mental model has been to think of all consumer software like slot machines. A slot machine is programmed to get a certain dopamine response from our brains. It's just an interface: a theme,

---

[*] J. Koetsier, "Flickr Founder Stewart Butterfield's New Slack Signed Up 8,000 Companies in 24 Hours," *VentureBeat*, August 16, 2013, https://venturebeat.com/business /flickr-founder-stewart-butterfields-new-slack-signed-up-8000-companies-in-24 -hours/.

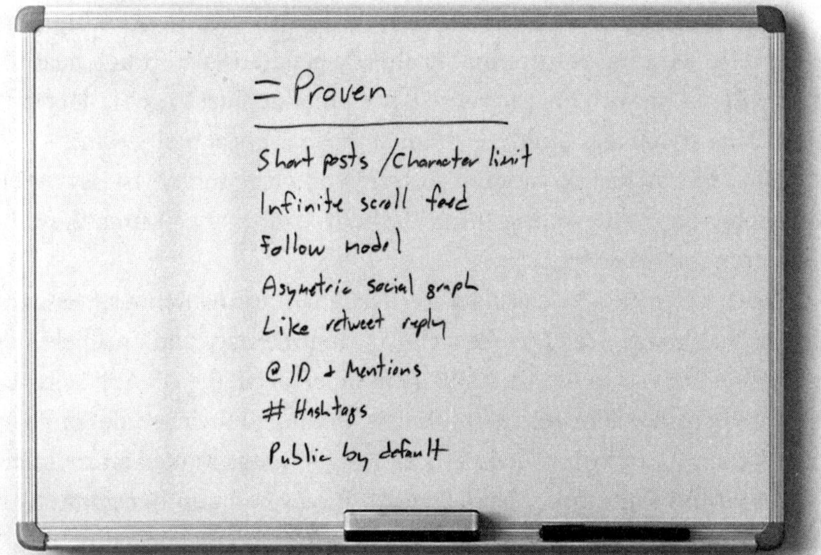

graphics, and tuning. If we built the exact same slot machine, we can imagine that users would have the exact same response.

For example, if you were the PM at Meta and Zuck asked you to deconstruct Twitter (aka X) to build the Threads product, what could you have identified?

I recently interviewed Nikita Bier, one of the most successful consumer app developers and now Head of Product at X, about his approach to Proven and finding heat. He defines heat as:

$$Heat = Frequency \times Intensity$$

Nikita measures this in Hourly Active Users (HAU). I love this concept of HAU! He points to Snapchat as the gold standard: "They took photos and made it a 20-times-a-day activity. They made photos into messaging, not just posting."

= Case Study: Nikita Bier—Finding Latent Demand

I first met Nikita when Zynga and Facebook competed to acquire his company, tbh (To Be Honest), which had virally expanded to be used by 20 million high school and college kids across America.

Nikita doesn't believe in user research. He believes in spotting *latent demand*—a winning feature that is already generating so much heat that users will put up with an otherwise flawed product just to get it. His job is to spot that signal and build the product those users actually want.

"Consumer is just so random in terms of what works," he says. "The best prototype is an existing thing in the market that captures the value you're trying to solve for."

In 2017, a teenager DMed Nikita about Sarahah, an anonymous messaging app that was going viral. Despite a clunky user interface and an Apple store description that was in Arabic, it still hit number one in the US App Store. Users shared positive anonymous comments, linking to their profiles on Snapchat. The app's virality revealed a crucial insight: Teens wanted affirmation.

Around the same time, Nikita's team discovered another game teens were playing on Snapchat that was just a screenshot of compliments—"I like your style," "You have great hair"—with emojis. The file was named tbh.png, and this became his inspiration.

Nikita saw a huge amount of latent demand, but no existing product was harnessing it yet. His version, tbh, was built to feel just anonymous enough, but with a better form factor. Instead of open text boxes, users voted on pre-written positive polls. (Staying close to the metal, Nikita wrote them all himself.) Nikita made a design choice to also add context to the poll responses: "This reply came from a 9th-grade girl." That small tweak made the messages feel real, not bot generated, and more "screenshot-able." It gave teens something they could post and say, "Who sent me this?" That turned out to be the magic loop to drive virality and constant engagement. In the first few weeks, tbh users shared 80 million screenshots of their compliments, nearly all via Snapchat.

The team didn't try to reinvent the wheel on anything else. They copied proven elements, such as contact sync from Facebook and Instagram and onboarding from Zenly, which had the highest conversion at that time.

Within a few weeks, tbh displaced Sarahah and took the number one spot in the App Store.

Years later, after his Facebook acquisition earnout ended, Nikita launched the same app again, only better. The top piece of feedback during tbh's first run was this: *Can I pay to find out who sent me the compliment?*

= Proven
Anonymous social
sharing
Posting screenshots
to Snapchat
> Better
Replace text
boxes w/polls
+ New
Positive only

This time, he built that feature, launched the app, and got to $11 million in revenue in 90 days—all without any venture money.

> Better Defined

Better is some aspect of the proven product that you can improve, so that 10 out of 10 users say "Fuck yeah" (or at least, click on the feature)! This isn't just your opinion: What we product makers think is Better is New! Better, however, is objectively better. Maybe it sells at half the price or is offered for free or doesn't have to be downloaded. In the case of tbh, this aspect was simply that Nikita's App Store listing wasn't in Arabic!

Better usually involves such small improvements that it's rarely enough for a product to win on its own. The primary exception is in a platform shift (discussed below).

Case Study: FarmVille

FarmVille was Zynga's biggest hit game, but, more important, it represented the tipping point at which social gaming reached mass-market adoption. The FarmVille franchise generated one billion installs and $1 billion in lifetime revenue, and it led to a broader mass market for gaming, which today generates globally around $200 billion annually with $35 billion identified specifically as social gaming. We built FarmVille in six weeks and launched it with just Proven and Better mechanics. So the game is a useful way to show how you can win without any New ideas at all.

For context, we had been looking for the right opportunity to offer free, social versions of simulation games with higher production values, such as *The Sims*, on Facebook. We saw this as a path to scale to a much bigger mass market in terms of both users and revenue.

One day, Justin Waldron (JW) ran into my office. "We need to buy this game YoWorld today! They've built *The Sims* in Flash!" JW was a 19-year-old game prodigy I had hired onto our founding team in 2007. I found him by fishing through the best failed games on Facebook. He had built a beautiful Flash game with no users. JW spent his first six months working on Zynga Poker from his mom's basement in Connecticut. We bought Yo World and relaunched it as YoVille under a new studio run by JW in May 2008. It was one of the first simulation games on Facebook.

At the same time, in late 2008, we saw the farm simulation category blowing up. A game in China called Happy Farm was exploding, with 23 million DAU, despite offering primitive art and only six fixed plots of land. In the fall, a UK developer launched MyFarm on Facebook. It proved a new mechanic—a movable, interactive game board that felt like an Etch A Sketch: You could click any pixel to place crops, giving players a large canvas for creativity. I was intrigued. A few months later, in early 2009, a Florida developer launched FarmTown, which became a viral hit. It combined MyFarm's game board with our YoVille player avatar and the neighbor bar from Restaurant City, a beautifully made game from our only serious competitor, Playfish (luckily for us, that company was later acquired by EA, which rendered them useless).

It's worth taking a minute to explain why Restaurant City was so innovative and how it accelerated its mass adoption of social gaming. First, it was one of the only non-Zynga games I was ever addicted to, which was a massive signal for me. I was so hooked that I got in a fight with my then-wife over using our only laptop on a cross-country flight because I needed to upgrade my restaurant. The two big innovations were a neighbor bar, which showed all your friends who were playing the game and offered the ability to "hire your friends" to work in your restaurant. (I think Playfish might have been a bit aggressive at first, because I saw someone had hired Zuck as their waiter, even though I was pretty sure he wasn't playing the game.)

On February 6, 2009, I wrote "Day 1" on the whiteboard in our e-staff room and told the team I would count the days until we were in the market with a farm game.

I always had a farm/ranch fantasy, but no one at Zynga thought it was cool, so everyone refused to work on the project even though the product was so clearly Proven. Even JW thought it was lame and instead built his dream game, CoasterVille, which unfortunately failed.

So, I acqui-hired My Mini Life, a four-person team that had built a failed Flash game. I matched them with Mark Skaggs, a talented, grizzled game maker with a background in role-playing games (RPGs). He was probably the last guy anyone would ever expect to make a farming and decorating game for middle-aged women. But I was determined. We added one artist and parked the group in an alcove outside my office, where we had a daily stand-up meeting.

We focused on making the farming core loop—plow, plant, and harvest—better because that was what players cared about. That meant better crop art, math, and UX design. Skaggs, who was a master craftsman, paid attention to the smallest details and noticed there were many unnecessary clicks in these games. He superimposed the right tool on the player's mouse cursor to have them avoid clicking through more menus. These were micro-polish moments.

There was another huge opportunity for Better. That January, I met up with Zuck for lunch. (Because Zynga was Facebook's biggest developer, Zuck and I had a monthly lunch to talk about the platform.) He told me, "We just opened the feed up to apps, but nobody's using it. Send me everything you've got, and I'll decide what to show people." We could make FarmVille the first game played in the Facebook newsfeed.

In FarmTown, the neighbor bar started out empty, so a player had to invite their friends to play by sending them Facebook requests. To Skaggs, that extra step looked like unnecessary friction, as most games at this point automatically showed friends who were also playing the game. So he got rid of it. Skaggs showed the game to JW, who had harsh but true feedback, scolding him for changing what was Proven.

Justin pulled Skaggs into the Mafia Wars tin room (which had one wall made of metal), where we had most of our product brainstorm meetings, and schooled him on Proven. It came down to my deciding vote. "Justin's right," I said. "We don't understand what's working and what's not yet, so we haven't earned the right to change it."

I said that we could A/B test and remove the extra step later, but for now, it would stay in. Skaggs was right that it was extra friction. But in this case, the empty bar had a massively important purpose: It drove a huge volume of friend requests, which increased the virality of the game and reminded people to come back.

Casual games are like a comic strip in the newspaper. None of us would go to a comic book store, but if you trip over one in your daily newspaper, you might read it every day. That's what casual games were like on Facebook. When users sent requests to their friends, it created a reminder to come back and play the game. If they had never gotten that request, we could have lost them forever.

We launched FarmVille on June 19, 2009, 133 days after I put the

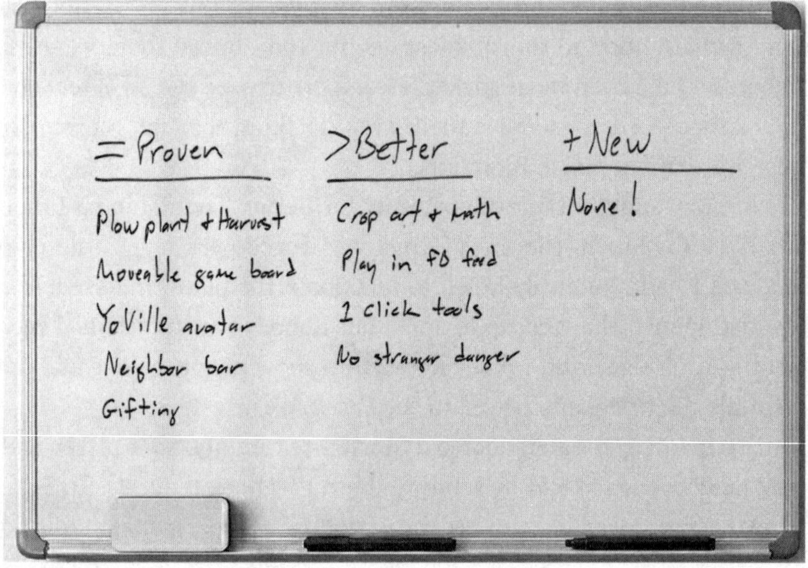

counter on the e-staff whiteboard. The game became an instant viral hit, reaching 1 million DAU within 4 days and 20 million within 6 months.

Case Study: Hay Day > FarmVille

One of the most successful examples of Proven Better (no New) came when a team at Supercell created Hay Day, which was essentially Farm-Ville for mobile.

If you compared the original Hay Day and FarmVille, you would have thought they were the same game. Both featured the same half driveway, the mailbox, and the house with the white picket fence. They even have the same pickup truck—one was red, one was blue.

But I'm not complaining. This is a master class in Proven Better New!

Ilkka Paananen and the team at Supercell had previously worked at a failed company called Digital Chocolate, which was started by the founder of EA, Trip Hawkins. They had developed a game called Millionaire City on Facebook but had failed to grow past one million DAU. I think Ilkka and the team were so beaten down that they were humble and curious. They were the best game designers in the world, and I admired them for applying Zynga's playbook to out-innovate us.

The irony is that the whole game industry had said that FarmVille wasn't even a game. Then, a talented team did a perfect Proven Better New of the very same farm game for mobile and crushed it.

+ New Defined

It's worth repeating my mantra: "All New fails." New is your novel idea that's never been tried with this particular product and audience. If you accept that this is most likely going to fail and therefore is the riskiest aspect of your product, then it becomes easy to isolate and test it separately.

Great product makers prioritize customer delight over growth; that means "don't change what already works." People don't like change. As product makers, we move faster by following repeat behaviors. So when some MBA changes the army figures in your Risk game to plastic Roman numerals, we're pissed. People were angry when *The New York Times* added color photos. It took Craig Newmark, a master product maker, two years just to add pictures to listings on Craigslist. A newly minted product manager (or, really, any of us) might have assumed everyone wants color images in their newspaper and pictures in their marketplace listings. Imagine driving across town to buy a couch you have never seen! But Craig, who has deep empathy for his users, understands that changes can be painful to the millions who rely on his service every day.

When pursuing New, we should be like scientists with white coats in a lab. We're trying to quickly re-create what's Proven so that we can generate what we believe is Better and prove we're right. Then we isolate the New variant, so that we can get accurate test results and not false signals. We're also isolating the New so that it fails for the right reason and doesn't sink the rest of the product.

+ *Case Study: Brian Grazer's Hit Empire*

Brian Grazer, the Hollywood legend behind *Empire*, has a simple playbook for making hits: You don't reinvent the wheel—you add New twists to Proven story tropes.

At its core, *Empire* was just another soap opera. It had the same mechanics, the same family power struggles, and the same Proven storytelling beats. But Grazer's New was one bold change: He shifted the cultural lens by centering the show on African American protagonists.

As Brian tells the story, the most important element in making *Empire* so successful was taking a Proven format and making it more culturally relevant to the music industry, while also giving diverse, historically underrepresented audiences representation on TV.

That one shift was enough to unlock something massive. *Empire* wasn't trying to rewrite TV; it was plugging into a format people already loved but reimagined through a different perspective.

In its first season, it became the number one watched show in the world. The premiere drew nearly 10 million viewers. The season finale attracted 17.6 million viewers, which made it Fox's highest-rated first-year series in a decade.

People chase New when they should build on what's Proven. Grazer's genius wasn't in breaking the system, it was in understanding it enough to make one change that could generate a hit.

Ask yourself: Are you building something people already love, just making it Better or New? Or are you trying to invent a product the world has never seen before?

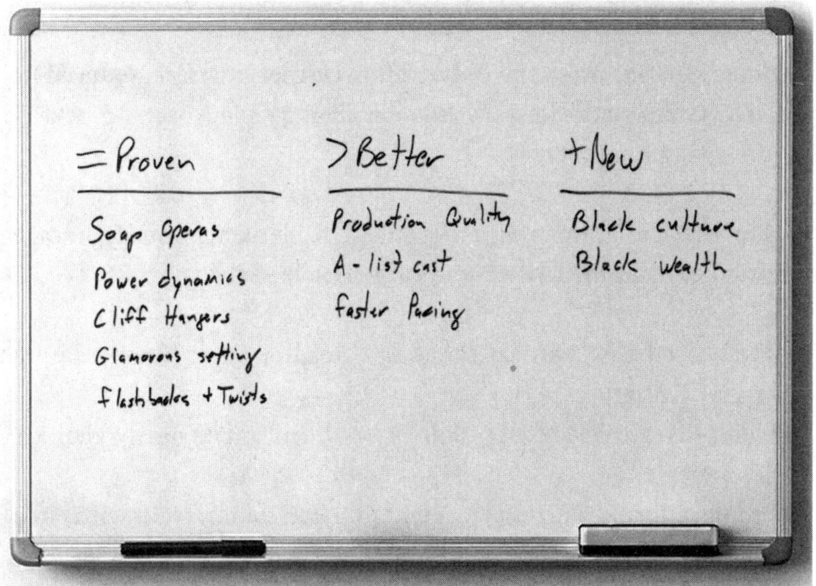

Exercise: Applying AI to Proven Better New

AI is an amazing tool to get us leverage in implementing Proven Better New.

We had a lot of fun uploading this chapter to ChatGPT and asking it to deliver a Proven Better New analysis of several hit products to explain their success—and then to help us generate several strong shots on goal. AI has already become a crucial and magical tool for product making, but so far, it can't replace what we humans add: creativity, empathy, and taste.

Case 1: Use GPT to Deconstruct a Hit Product—Threads Example

We asked ChatGPT to show how Meta's Threads product—a fast follow of Twitter/X exemplified Proven Better New—became a record-breaking hit, reaching 100 million users in its first five days after launch.

This exercise was a magical unlock moment for me—I saw a glimpse of the future of product making. The AI got to such granular detail that it gave me an almost-buildable product spec.

Threads Proven Better New:

= Proven: Instagram + Twitter mechanics (feed, interactions, visual-first UX)

> Better: Instagram-graph onboarding, cleaner interface, reduced noise

+ New: Cross-platform growth loops, identity piggybacking, text layer on Instagram graph

= **Proven.** These are mechanics and UX elements already known to work from Instagram, Twitter, and early mobile social apps:

= Feed-based UX: Vertical scrolling thread of posts (Proven by Instagram, Twitter).

= Heart (like), reply, repost, share icons (identical or nearly identical to Twitter).

= Profile ring and stories placement (top horizontal layout with circular profile pics—borrowed from Instagram).

= Camera-first design for quick creation (Snapchat-inspired).

= Engagement loops such as likes and comments (from existing social platforms).

= Push notifications and follows for virality (standard across social apps).

= Threaded conversations as seen in Twitter's reply stack.

> **Better.** These are user-facing improvements where the UX or performance feels superior:

> Cleaner, less noisy UI: Threads strips away most clutter from Twitter/X—no trending hashtags, fewer promoted posts.

> Instagram-integrated identity: Eliminates the need to rebuild your graph. You start with your followers already there.

> Onboarding is instant: Log in with Instagram, import handle, bio, follows. Feels 10 times faster than Twitter onboarding.

> More friendly and controlled feed: Prioritizes known accounts, reduces algorithmic chaos, especially early on.

> No ads (initially): Creates a premium feel, even if temporary.

> More visual-first styling: Better rendering for links, images, and text formatting.

+ **New.** Here's where Threads isolated risk and tested novel ideas:

+ Instagram-native follow graph: First app to fully piggyback a social graph across Meta's ecosystem in a new app. This was never done before at scale.

+ Cross-platform engagement loop: Posting in Threads nudges notifications on Instagram and vice versa, fueling cross-app growth.

Case 2: Use GPT to Build Your Own Hit—Can We Beat Threads?

While AI did a strong job deconstructing why Threads was a hit, it fell short at helping us brainstorm a product that could beat it in the market today. You can see the Proven Better New analysis in Appendix II. Here, I'll sum up how it did:

- **= Proven.** GPT gets a B+ for this. It nailed the key points about core UX and growth and virality mechanics but missed that using people's existing social graph offered instant relevant content from people you cared about. This was a new idea that Threads proved worked.
- **> Better.** GPT gets a C on this. Every feature it suggested, such as feed control and conversation tools, were New ideas. GPT got the concept wrong, but its Better features were pretty good New ideas to test. Interestingly, this is the exact mistake I see most people make: They miss seeing that what they think is Better is really New.
- **+ New.** GPT gets a D on this. While it tried to recommend some novel ideas, such as rooms and live simultaneous posting, they were surprisingly conventional.

GPT is great at looking backward to analyze and explain the world around us. But as of today, it doesn't have the ability to distinguish between false and true signal; it can't ask 10 users what they'd say "Fuck yeah!" to; and it has yet to match the human ability to predict and imagine new products, new markets, and new worlds. Luckily, there's still a big role for us humans to play in product making, and as we've just seen, AI can accelerate the work of the best product makers.

Where to Apply Proven Better New

Platform Shifts: The Ultimate Opportunity

The best place to apply Proven Better New is during a platform change. When platform shifts occur—as we saw with web, then mobile, and now AI—people are in discovery mode. This means the barrier for entry is low, and user acquisition is cheap, even viral.

This concept has played out in the game industry over decades. When the next hardware version of Xbox or PlayStation came out, game companies could reliably invest in graphics that would use that chipset better.

Then, they could release a more stunning version of *Grand Theft Auto* or *Madden Football*, knowing that their loyal fan base would be excited to buy it.

When that game platform changed, there was also an opportunity for someone else to steal a franchise. With the advent of the Xbox 360 and PlayStation 3, *Call of Duty* displaced *Medal of Honor* to become the top-selling World War II first-person shooter game.

While there's a moment of vulnerability when the platform changes, the incumbent still has an advantage, because there are still fans attached to the brand.

Zynga leveraged Facebook, the web, and Flash to capture mass-market adults, previously considered an impossible (or even nonexistent) market. Supercell leveraged the move to mobile, which the company still dominates. In a New platform, with users all in discovery mode, you can offer Proven Better and no New. What's New is the platform itself.

Today, with AI, it feels as if every market and incumbent were up for grabs. A great recent example has been Perplexity taking market share from Google. That outcome seemed impossible for the previous 25 years. I can't keep up with the number of start-ups hitting $50 million annual recurring revenue in every vertical sector, from legal to accounting to construction.

Mature Markets Require Finding a New Vein

The opposite is true in mature markets, where you're trying to find a vein. People are already committed to services in every established category, so you need a huge amount of New just to get them to consider and try something.

You have to pursue truly novel ideas. You need to get to what Bing Gordon calls OMFG moments. The back-of-the-box selling points need to be more extreme because the space is crowded. In a red ocean, a lot of people are going after the same users, who are no longer in an easy discovery mode.

A corollary to this is that mature markets usually lack innovation because they're dominated by incumbents that don't need to try risky new ideas. These are markets that the industry assumes are over but that sometimes haven't even started to see their real growth curve yet. Two examples of cases like that, where people thought the market was saturated but in fact growth had barely begun, are search before Google and games before Zynga. In both cases, the new entrant innovated on the core value of the mature market to unlock massive new growth. Before Google, search was a multibillion-dollar mature industry with low growth. Google made search so fast and relevant that our use cases expanded. GPT is now reinventing search and again expanding the use cases and market. Zynga made games accessible, fast, and productive, so much so that busy adults could justify playing them. I can't overstate the power of these kinds of innovations in mature markets. The users and business are already established, and you might turn billions into trillions.

Today, we all sense that AI will turn search and every major consumer and enterprise category into growth markets again. We've seen Chat GPT reignite consumer interest in searching and even prompting. And we've seen Perplexity get billions in venture funding to compete with Google, when only a few years ago, no VC would have touched a new search entrant.

Don't Bet on Wildcat Drilling

Whether you're analyzing investments or formulating your own product strategy, Proven Better New can derisk your approach. Rather than pursue *wildcat drilling*, or the equivalent of oil companies randomly picking new holes to drill, companies that systematically apply Proven Better New offer far better risk-adjusted returns.

I created a two-by-two matrix to help my Stanford MBA students see why just pursuing New ideas without considering what's Proven is fool's gold. In contrast, companies that pursued Proven Better New have far greater odds of success. The grid compares the relative risk and margin potential of different product strategies in the game industry, but we can apply this same framework to any consumer industry. (MBAs love frameworks!)

RISK

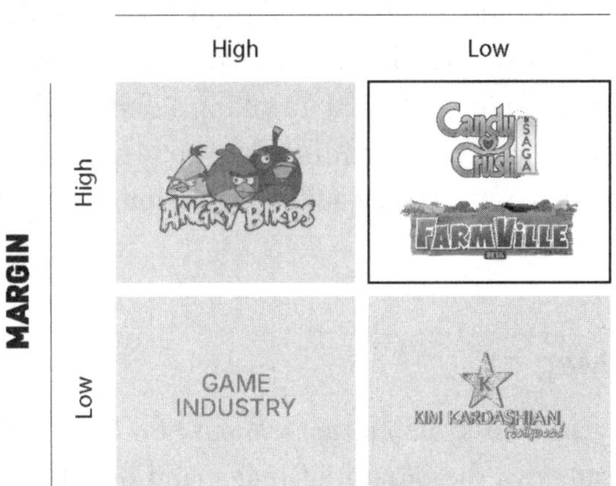

1. Fool's Gold. The upper left in the matrix is fool's gold: One wildcat driller found oil, but in a way that's not repeatable or predictable. It's "Newville"—trying to do things that are new, but not likely to work. That's Angry Birds. It was Rovio's 45th shot on goal—where every shot was different.

2. The lower-left quadrant (the majority of the game industry) is high risk, low profit. That's wildcat drilling: You're just trying new things with no evidence they'll work.

3. The lower-right quadrant is low risk but low margin. You're leveraging intellectual property that's so proven to drive traffic that the IP owner makes all the margin. That's also fool's gold.

4. The upper right—high profit margin, low risk—is where Proven Better New lives. That's where you want to be. FarmVille, Candy Crush Saga, Hay Day, and Meta's Threads were all good bets and had great odds of success.

When Proven Better New Falls Short

Proven Better New is so powerful that it can become a crutch: It can deliver you false signal in the form of B+ products that never get to A+. Because you are copying successful products, you may get positive signals, even though your unique new idea isn't that good.

AI and LLMs offer the potential to massively accelerate Proven Better New, but are also likely to homogenize, reducing both quirkiness and the potential to surprise and delight people. GPT can get you to a B+ in seconds, but it might actually be pulling you away from getting to an A.

That's kind of like what happened with the writing of this book. I tried to use AI to write it faster, but it didn't sound right. AI can be the enemy of quality: It flattens your voice, strips out the nuance, and turns everything into soulless, derivative soup. We ended up taking a nonscalable approach. The process was long and at times painful, but you're hearing my human voice. That voice comes out often by adding more long-form stories that capture my lived experience.

The Pursuit of Magic

I've spoken with Brian Chesky about this—about how Proven Better New needs to be balanced with the pursuit of magic and quality. Sometimes, as with Brian's launch of Airbnb Icons, there needs to be a Steve Jobs–like pursuit of something just for the sake of magic.

In 2024, Brian launched Icons, which are once-in-a-lifetime travel experiences. They included sleeping in a leather bed in the Ferrari Museum in Italy and doing a hot lap in the Ferrari 296 GTB the next morning, drifting off in Disney and Pixar's Up House (yes, it floats), and hanging out with Kevin Hart in a secret speakeasy.

In fact, Brian made a point to share his Icons product with me early, because we both knew no one had PMed or optimized this. I loved the magic but questioned why he didn't add more around it. My feedback to him: "Every time you get an Airbnb reservation, you should have a chance at a golden ticket!"

But Brian, in a very Jobsian way, was focused on quality for the sake of quality. And there's a magic in that—to offer people amazing moments and not ask anything from them. To just deliver value but not try to get all the meat off the bone—no marketing value, no revenue. It was just "fall in love with our brand and fall in love with travel again."

It's not clear what value Airbnb got from this launch, but I love that he pursued it. In many ways, he seems to be elevating our craft, giving hope, not just to his users but also to his peers, that we can pursue magic before metrics.

I'm torn between my admiration for people who pursue magic and quality for its own sake and my PM instincts, which tell me we can't just build it and hope they'll come. Maybe the answer is that there's a time for scrappiness like when Brian was starting Airbnb, and later if you achieve big success, you have more room to pursue experimentation. Maybe we can just think of this as real R&D for product making.

TL;DR

A core philosophical question about product development is whether it is an art or a science. My answer is that it's both: The art is taste, and the science is process. The best product makers don't chase novelty for its own sake; they start with what's been proven, they build with discipline, they innovate where it counts, and they know when to break the rules to deliver something magical.

5

Roadmapping Is Your OS

The best way to operationalize Proven Better New is through *roadmapping*. This practice is your operating system; it drives everything in your company. It's not just a weekly meeting—it's the most important hour of the week.

Your roadmap is where strategy meets execution. It's a living document that defines projects by resources, outcomes, and dates, and sets the pace and priority for your whole company. It's combined with a regular (for me, weekly) meeting. It's the tool that determines which hills we take, what outcomes we expect, and how much engineering effort each project is going to cost. It's a living weekly report card.

At Zynga, roadmapping was how we ran our factory—our product-making process from design to engineering. It allowed us to translate raw heat from ideas into features that had clear, measurable player outcomes. Funnily enough, I didn't do roadmapping at all before Zynga. I was driven to it because we needed to learn faster. We were in a sea of chaos and needed to stop time, which we did by creating and following this intellectual discipline.

If nothing else, as a product organization, you should be getting better and more disciplined with every release cycle. As my tennis coach used to say, "Don't play to win. Play to get better. I don't care if you hit the ball out of the court. I just want to see you commit to a proper swing."

Roadmapping also serves as a weekly diagnostic tool. It's a quick way to tell us whether the gears are connecting. In other words, are engineers coding features people actually want? Are we learning fast enough? At Zynga, we were testing more ideas in a week than our competitors tested in a year. We needed these meetings to make sure we were making what our players wanted.

Most teams will try to cancel roadmap meetings. They'll say, "We're not ready. We don't have anything to show you." That's exactly when you need the meetings the most. Your roadmap is a mirror, and sometimes what you see in it is not pretty.

To be clear about what's at stake, here's my rule of thumb: When you get your roadmap right, you can make 80 percent of your engineering days shots on goal—features your users may actually want. The best of your competitors are probably running at 20 percent. So you could be operating at 400 percent of their efficiency.

Elon is known for asking, "What did you get done last week?" That's exactly what a roadmap is about. What has your factory produced? What is it producing this week? And what are you committing to for the next development cycle?

The need for a process such as roadmapping really hit home before I ever created my first roadmap.

My Roadmapping Origin

In October 2007, Zynga Poker had 50,000 DAU, and we were fighting for our life every day on Facebook. Imagine that your users have to consciously navigate deep inside another website to load your app every day, or you're out of business. We were a patient with such a thin pulse that any interruption in Facebook usage patterns would kill us. In fact, the Christmas break of 2007 killed Jetman, which went from the number one Facebook app to gone in two weeks.

One Monday during that October, I had a meeting that determined the course of Zynga's future. It was our first roadmap meeting, even if we didn't have a name for it yet. My CTO, Mike Luxton, told me we had to stop all development on Zynga Poker and spend the next two weeks rewriting our SmartFoxServer code, which couldn't handle our number of concurrent users.

We were redlining on two fronts: uptime and maintaining user engagement. I faced an impossible trade-off in booking my small, limited factory. I had to decide between risking our game crashing, which could mean instant death, or a slower death of fewer people showing up each day if we didn't add any growth features. I chose to keep building features to spark more player engagement, because I figured if nobody showed up, it didn't matter if we had a game available. But the worst of all worlds would have been to burn two weeks on a feature nobody wanted while also failing to fix our unstable game code. So I had to be right about what to build.

We bet it all on a feature, weekly team tournaments, that let you and your friends compete with other teams by combining weekly winnings for big chip prizes. Luckily this generated big heat, and our servers didn't crash. In the following two months, we quadrupled our audience to 200,000 DAU.

This meeting was so impactful that we started to have it every week. It became the single time when we considered all ideas, negotiated trade-offs, and lived with the consequences for another whole week until we fought things out again. Patterns quickly emerged. If engineering committed to building something and missed, there was nowhere to hide. Similarly, there was nowhere to hide if I demanded a big sprint for a wild idea that failed to deliver. I soon realized this was my real board meeting—where the most important decisions that affected the direction of the company were made.

This roadmap meeting became the core engine of everything we did at Zynga. Eventually we codified all of our objectives and competing projects into one meticulously organized document, which became known as the *Blue Sheets*. The Blue Sheets became the basis for a weekly meeting that was the single most important hour of our week. Eventually, these one-hour meetings took up half of my week as we grew and launched many games and studios.

Business Outcomes

QA Goal	Current	Remaining
20,000,000	19,225,852	-774,148.00
$78,000,000.00	$75,679,000	-$2,321,000.00
4.5	4.4	-0.1

Release (Live Date)	Feature	Expected Business Outcomes	Actual Business Outcomes	Dev Cost (# of Days)
Q4 Week 6 (2 Releases back)				
	Decoration Storage	1m DAU sourced, +$80k revenue	+670K DAU, +$60K revenue	30
	Mutual Friends Neighbor Invite	+50K Installs/day	**Deprioritized - negative impact on performance**	17
	Neighbor Visit Reward Bonus	+150bp Increase in DAU Sending Neighbor Requests (100K DAU Sustained)	Released as 20% Test. Increased D1 retention by 23 bp, Increased D7 retention by 15 bp.	6
	Dynamic Neighbor Gated Quests	50k installs/day sustained	+61K Installs/Day	14
	Quests 50-60	$40k / day	Deprioritized due to higher priority features.	3
	Content Release	$60k Spike	$75k	6
	DAU Forecast (7 day avg)	18,290,050	*18,386,970*	
	Gross Revenue/Day	$800,000.00	*$856,000.00*	
Q4 Week 7 (1 Release back)				
	FTUE Optimizations (Suppress gift interstitial for new users, auto-neighbor)	+20 bp D7 retention	+30 bp D5 retention in experiment	7
	Holiday Celebration	+1.2mm DAU/Day	+1.2mm DAU/Day, +1.6M DAU spike, 800K DAU steady state	45
	Instant Grow Crops (Cash)	$150k spike/$50k sustained	+$140k spike, $80k steady state	15
	DAU Forecast	18,665,917	*18,660,240*	
	Gross Revenue	$858632.18	*$791,703.00*	
Q4 Week 8 (This week)				
	Non-App Crew Buildings (Visitor Center)	**71K install/Day Spike, +35K installs/day**		22
	Autho Add 3 neighbors on install	+20 bp D1 retention (Test)		7
	Gift Interstitial showing friends that just started playing	+20 bp D1 retention (Test)		4
	ZLoc - Indonesia	491K DAU over 3 months		8
	DAU Forecast	19,144,394		
	Gross Revenue	$838,991.00		
Q1 Week 9 (Next Release)				
	Factories 1.0 [BB]	**1.4M Spike DAU, 400K DAU Steady State**		60

The Blue Sheets

By the summer of 2008, we had multiple product teams pursuing different games across four social networks—Myspace, Bebo, Tagged, and Facebook—each working from their own analytics and roadmaps. We were operating like several small start-ups. Teams were repeating the same mistakes. They were never sharing insights or, more important, breakthroughs. I found myself constantly pushing the teams to talk to one another, but nobody had the time. We may have been testing more ideas than anyone else, but that knowledge was siloed, buried, and useless. I made a top-down call that changed everything: "Everyone is going to work off the same roadmap document." It became known as the Blue Sheets. I have no idea where this name came from other than someone chose to make them blue and it stuck.

At first, teams fought the change. They acted as if I were a CEO out of the *Dilbert* comic, forcing them to be corporate rather than fast. They kept two sets of roadmaps: one they worked from and one to "communicate up to Mark." This lasted for around three months. I was starting to question my decision, too. But then, one week, a team found a viral breakthrough, which everyone else saw and copied immediately because we all shared one analytics and reporting system. Once they saw another team nail something and immediately put it to work themselves, the game changed. This was what turned the Blue Sheets into a competitive advantage—it enabled us to adapt and optimize improvements between and across teams.

Zynga's Roadmapping Process

Roadmapping at Zynga empowered GMs and PMs to operate as CEOs, each running their own factories with full P&L (profit and loss) responsibility. They quickly realized that making the right weekly decisions could directly improve both player and business metrics. Over time, this roadmapping evolved into a complete operating system through which we ran the company, which grew from a small team to thousands of employees and booked billions in revenue.

Roadmap meetings enabled me to personally drive urgency and accountability throughout the whole company every week. By anchoring on our Blue Sheets, I could hold teams accountable to deliver on what they said they would and push for big ideas and growth in the near term.

Regardless of how big we got, these meetings let me stay close to the metal. I could check in on what we were building, see what was working or not working, and track whether our teams were pursuing big ideas. The meetings also enabled me to quickly share best ideas and practices across game studios every week.

Studio Priorites

- (P1.1) 7M by EOY - Executing on Growth Roadmap
- (P1.2) Currency + Sinks - Hindsight launch in September
- (P1.3) New Game feature set

OKRs

KR1 - Reach: 7M by EOY (6.7M EOQ) (7/10)

KR2 - Revenue: Revenue target: $44.3M (9/10)

KR3 - Bold Beat: 80% of DAU us currency + sinks before EOY (7/10)

KR4 - Growth: New Game SL on Androidt OS by EOY (8/10)

App rating: 4.5/5 (10/10)

Projected Weekly Growth: 0.2%

Projected Weekly Rev Growth: 5%

This Month's Objectives

- Executing on goal of 7M DAU by EOY
- Set final product scope New Game
- Multiple Weekly Challenges launch
- What will our player thank us for — First Live stream went live last week with great community feedback

Team Health

Current team health   Green

Commentary on team health

- Team focused audience growth and hitting production milestones
- Starting to see momentum in Growth metrics and stability in DAU

Roadmap meetings started with reviewing the "four quadrants," which captured studio priorities, ***objectives and key results*** (OKRs), and team health (green/yellow/red). Were teams on track or spinning? Then, we dove into the Blue Sheets, which drove the meetings. If the Blue Sheets didn't show growth, I would ask them to repeat the meeting later in the week. The pattern recognition I quickly built was that teams that lacked big ideas were drowning in small features we called ***mouse nuts.***

The atomic unit of the Blue Sheets was usually a new feature, although they often included many other necessary projects, like content drops, which were usually new additions of game art. The image below shows a typical feature analysis.

Feature	Expected Business Outcomes	Actual Business Outcomes	Dev Cost (# Days)
Holiday Celebration	+1.2mm DAU/Day	+1.6M DAU spike, 800K DAU steady state	45
Instant Grow Crops (Cash)	$150k spike/$50k sustained	$140k spike/$80k sustained	15

The real power of roadmapping wasn't just in the accountability it made possible; its real power was in how quickly teams could learn from each other. When the FarmVille team discovered that players who helped neighbors water crops had three times better retention, every other team began testing their own version of the feature within days. Ideas that worked spread Zynga-wide within hours. We could do Proven Better New against our own games every week. Turns out nobody feels bad about copying inside the same company.

The system fed itself. Teams that hit their metrics earned more resources, which meant more testing, which led to better metrics, which got them more resources. But here's what made it really work: We never mandated how teams should hit their numbers. The system rewarded outcomes, not process.

Every studio created a war room, covering the walls with idea funnels and Proven game mechanics from previous weeks. These rooms became our nerve centers.

Test More Ideas in a Week Than Your Industry Does in a Year

This was a mantra inside Zynga and still is for me today. Since all new ideas fail, we want to build failure machines. Imagine if your team could fail 1,000 times a week while your competitors are failing once a year. The point is that eventually one idea won't fail.

At Zynga, we tested ideas with links and pop-ups in our games. We eventually built our own ad-serving tech to test 100s of ideas a day. Today everyone can run tests on ad networks and even virtually with AI agents that can simulate our customers. Ethan Mollick, who runs an AI lab at Wharton that I've funded, has shown that AI tools allow product makers to test new features 10 times faster.

Build a Teaching Hospital

A great way to build an organization that can learn and adapt faster than anyone else is to build a *teaching hospital*. Similar to a surgical theater in a medical school. I operated on every game in a windowless room surrounded by moving whiteboards. The team sat at the table in the middle, and new hires and other Zynga employees sat around the perimeter listening. (Most of these meetings took place in a room someone found at the bottom of our De Haro Street building, which, according to company lore, had once been a DVD porn shop. Now it's Discord's HQ.)

At Zynga, our teaching hospital was about building an organization that could learn and adapt faster than anyone else. The combination of our ruthless metrics tracking through Blue Sheets and our teaching culture turned us into a living laboratory, where every feature launch was a chance to get smarter as a company.

My three objectives for every roadmap meeting were to ensure:

1. The game is tracking to its OKRs.
2. The team is being creatively productive and operating efficiently.
3. New product leaders are leveling up.

Every game studio had to show up with the following:

- A hypothesis they were testing that was tied to their objectives
- *Expected outcomes* (EOs) per engineering day (straight from their Blue Sheets)
- Real metrics versus expected
- What they learned and how it would shape next week's bets

It was intense. If you didn't know your numbers, if you didn't have answers, or if you weren't signing up for growth, I'd say, "Go back and do your homework again. We'll meet again later this week or on the weekend."

Our Friday happy hours weren't just for letting off steam; they were part of the teaching hospital. We showcased and celebrated the ones who hit it that week: Who had a big release? Which feature moved the needle? What did we learn that could help other teams?

This worked because of how Blue Sheets fed into our teaching culture. If a team's Blue Sheet showed a win, the team didn't just celebrate; they'd teach. They'd break down how they spotted the opportunity, how they estimated engineering costs, which metrics they tracked and why, and how they iterated based on the results.

Other teams didn't just copy the feature; they learned the thinking behind it. This way of working allowed our success patterns to spread rapidly, but more important, it created a culture in which teams learned quickly about product development and innovation.

Every successful feature became a case study. For example, when Words With Friends saw weekly challenges generate 40 percent more games played, the team didn't just high-five over the metrics. They broke down exactly how they designed the feature for max engagement and what surprised them in the data from the other game studios.

OKRs: Know Your Goal or Suffer Death by 1,000 Bad Features

The driving force behind our roadmap meetings was our commitment to OKRs. We used this system to define the level of growth, risk, and innovation we expected every team to sign up for every quarter. OKRs worked hand in hand with our roadmaps and roadmap meetings. The OKRs were literally listed at the top of the roadmap document, which showed weekly plans to achieve progress.

The process to get to OKRs was a quarterly top-down, bottom-up negotiation with our teams. For me, this usually started with an expectation of 30 percent quarterly growth because I thought that was possible and it was a message to the teams and a reminder to myself that I wanted us to go for it every quarter and not pursue incremental ideas.

OKRs are a system that forces product teams to prioritize a single objective, supported by several key supporting results. I gave a three-hour lecture on OKRs at Harvard Business School, which I titled "A Love Letter from the CEO." OKRs invite innovation and autonomy, while also driving accountability and growth. They also give the CEO a framework to manage and course correct, whether that's with one leader, one product, or the whole company.

OKRs have been controversial in Silicon Valley as they've often been implemented badly, losing the plot to the extent that they become *Dilbert* comics—fake and corporate rather than actually useful for teams. You will have to decide in your own particular business and team whether

OKRs are valuable. Testing for one quarter could be a good, low-risk way to do so.

I've found OKRs most valuable in keeping teams focused on big innovation zones. The biggest trap most product teams fall into is that they launch many mouse nuts features, which deliver short-term results but make their products worse over time. Today, so many mobile apps, especially games, force users to search for the x to close all the pop-ups, as if they were playing whack-a-mole. This occurs because teams lose sight of the goal, what they're innovating against, and in its place they chase caffeine-hit features. The worst lately is the spin-the-wheel pop-up on every website and app. To gamify a product doesn't mean to literally add a random game to your app! My daughters spent last summer redoing the first-time user experience (FTUE) for Pokémon Go. They advised removing most of the pop-ups.

In Silicon Valley, there is both a religion of OKRs and hatred for them. Andy Grove first invented OKRs at Intel as a way to drive accountability across his entire organization toward big chip launches that would decide the company's future. He needed every division to deliver its piece to hit the launch date, so that the company could ship the next generation of PCs based on the next faster processing chip: 286, 386, 486.

Legendary venture capitalist John Doerr spread this gospel to the rest of the Valley, most famously while at Google. When John came to the office and evangelized OKRs at Zynga, we became true believers. By 2008, Zynga had exploded to 200 people with more than 12 games across 6 offices. We needed a process. Our decentralized studios, built around clear daily metrics, mapped perfectly to Intel's original OKR approach. We could easily measure each studio's effectiveness, normalize the data across the organization, and directly tie engineering days to player metrics and revenue.

Implementing OKRs led to multiple benefits for all of Zynga. Most important, it enabled me to put a flag in the ground every quarter and communicate to the whole company that we were signing up for bold, ambitious growth—and all the risk that came with it. We arrived at studio OKRs through a top-down, bottom-up negotiation, and most of the time, the OKRs started with a 30 percent quarter-over-quarter growth objective.

This metric communicated to teams that they couldn't keep doing whatever worked last quarter. We needed to chase big ideas constantly to grow our player base.

Nobody was allowed to change OKRs mid-quarter, even as the gaps became more evident as weeks progressed. That rule forced us to constantly rethink our execution path and often take on greater risks as we got further from the goal. This is the opposite of how most teams operate: They become more risk averse and pursue smaller, "sure thing" features that actually make most products worse over time.

I want to acknowledge that while this system worked perfectly for us, it has failed miserably inside many other organizations for good and bad reasons. OKRs might not work for a product company that may be organized functionally, like Apple or Airbnb, rather than in decentralized product teams with their own P&Ls. Similarly, a single founder and team working to get from zero to one may find the model to involve too much overhead.

I find OKRs work better for live products than for new ones, so I use OKRs more loosely in the beginning, when it's just me and a single team trying to launch our first product. Post-launch, once you have multiple products or product initiatives, OKRs can drive more intellectual honesty and accountability.

Use OKRs to Chase Audacious Goals

We can use OKRs to turn our boldest dream into a measurable, achievable reality. I call this "the hill we are taking."

In 2007, I put a giant sticky note on my wall with the most audacious goal I could imagine us achieving in the next three years—to build the YouTube of games. Each box had a goal along with audience size. At the time, building the top branded game network on the web, with 100 million players, was the biggest goal I could imagine. In fact, we beat each of those numbers by 15 percent, but I'm still amazed how accurate that note turned out to be. The point was that having a goal defined the mission.

To achieve greatness, we have to hold ourselves accountable to what's theoretically possible, either because we've seen others do it or because we know it's within the laws of physics. That's how we get to great outcomes—

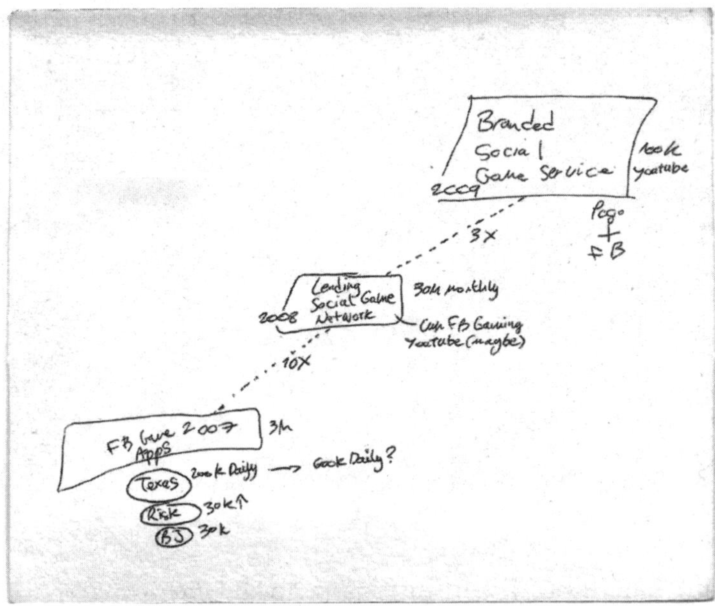

not by incrementing from where we are, and not by being rational and reasonable.

Great leaders don't just make up a wild goal and call it a moonshot. Elon always asks, "Is it within the laws of physics?" I often held teams accountable to what we saw our best competitors achieve in the market. I thought of it as like a cyclist pulling away from the Peloton. We had to draft off their momentum, catch them, and then pass them. A great example was in 2010 when Digital Chocolate published Millionaire City, which achieved 30 percent DAU/MAU. We had never seen that level of user engagement before and it immediately became the new benchmark for all our teams.

What Winning Looks Like in OKRs

The accompanying image shows our Zynga company-wide OKR review for 2009, when we were firing on all cylinders and everything was going right. Our objective was to achieve "escape velocity." And we clearly accomplished this. We had 9 of the top 10 social games and were as big as our next 11 competitors combined. Our games had become household names.

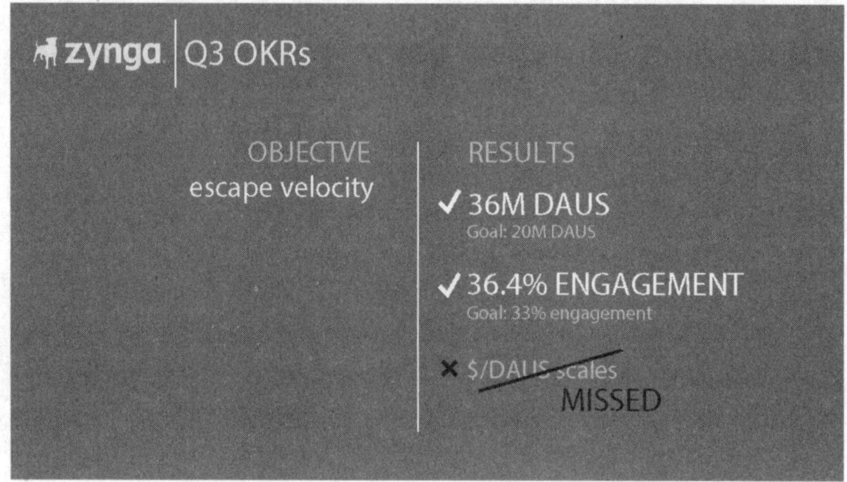

Invent your own metrics

Too many founders work with high-level standard metrics like DAU, retention, Average Revenue Per User (ARPU). The problem with these metrics is that they are not leading indicators; they're trailing outcomes. So, for every product we need to organically identify the metrics that are both early indicators of the future success of our product and ideally represent levers that we can move. In the case of Zynga Poker, we obsessed over hands played and in Words With Friends turns taken because both showed us the true indication of our player engagement, and these also turned out to be the right metrics for our teams to focus on moving.

Facebook famously had a magic number of seven friends, after which point they knew that they had us all for life. This metric made it simple for their teams to operationalize and work and test. That inspired us to think about the right metric to measure our Cocktail Party. How could we tell if a user liked it? And how would we measure if they wanted to stay?

That's what ***Active Social Network*** (ASN) became for us. ASN measured the number of round trips players had with friends (e.g., how many turns taken in Words With Friends or number of gifts sent and received in CityVille). We found that when a user went from 0 to just 1 ASN, there was an 80 percent chance we'd see them again in the next month. When a user went from 1 to 4 ASN, there was an 80 percent chance we'd see them 25 out of the next 30 days.

ASN was a perfect metric for us that created a virtuous cycle that motivated our teams to drive more social interactions across all of our games since we saw measurable impact on all business metrics.

Conversely, when we focused on improving our Net Promoter Score (NPS), it was a total disaster and demoralizing for our teams. We couldn't move the metric at all. NPS is a simple measure of your customer's likelihood to recommend your product to their friends on a scale of 1 to 100 where 1 means they're a detractor and 100 means they're a superfan. Bing, who was then on the Amazon board, sold us on using NPS as a measure of our brand value. He pointed out that the top consumer brands on the internet like Amazon, eBay, and Google all had a 70 to 80 NPS, which we saw as the gold standard for an Internet Treasure. We found that FarmVille had a 35 NPS, which was surprisingly similar to Facebook itself, whose NPS was 35 to 45.

We spent two quarters pursuing OKRs focused on doubling our NPS to the 70 range. The problem was NPS is a trailing metric, not a leading one, and we found it impossible to move directly. My lesson here was once you have metrics your team can't move, they ignore them. They focus on the metric they can move, which usually defaults to revenue. Cadir and I used to call this *squeezing the lemon*—optimizing for monetization rather than focusing on user experience or bold new features. We can all see painful examples of this today where apps we used to love bombard us with pop-ups with the close button harder and harder to find or YouTube ads that make us wait 15 seconds for a skip button.

We should always be looking for metrics that are both specific to our products and leading indicators for user engagement. It's important we pick the leading indicators even beyond engagement that really represent our players getting real value out of our product, and that should lead to higher NPS over time. We have to beware of leading indicator metrics that result in local optimization but long-term negatives for our players. One example was in the Zynga game YoVille where we created a bakery. Our players had to come back and work in the bakery every day, and the next day they could harvest a big coin output. It increased our short-term engagement but reduced long-term retention because it became a job and wasn't fun.

Easter Egg

Focus on Day 365 Retention (Nobody Else Does)

The single most important metric to show the enduring value of our product is ***Day 365 (D365) retention***. D365 retention measures how many people using our product today used it on this day a year ago. The Internet Treasure services we rely on every day like Google and Amazon probably have a D365 of 80 percent, which means that eventually everybody on the planet will use their service. Whereas other durable long-term services like LinkedIn or Snapchat by my estimate have a 30 percent D365, which means that they will potentially have a third of people on the planet using their service. That is still massive. By comparison, most games have a D365 of zero. However, we can see that some games—*World of Warcraft*, Candy Crush Saga, or Zynga Words With Friends—have achieved by my estimate a D365 of 10 percent, which is high enough to sustain durable multibillion-dollar franchises.

Based on this insight, I was a long-term shareholder in Twitter, even though they were hated by Wall Street, because I believed that I would be using them as my primary source for news in 10 years, and I still believe that today. Similarly, I've been a long-term shareholder in Snapchat, which is also hated by Wall Street. Even though I don't use it, I see its long-term durability with my kids and their generation. I also believe that network effects once created are nearly impossible to destroy. It's amazing that the Facebook blue app is still massive even though nobody I know has used it in years.

As product makers, D365 is a powerful North Star. It's useful for us to imagine whether our users will still engage with this in a year when we're considering a new product or feature. Even though, similar to NPS, D365 is a trailing indicator, it's still one of the most powerful metrics to show over time whether we are building long-term value and enduring brands.

Aspire to Run A+ Roadmap Meetings

The first time I met Bing, he sat in on my Poker roadmap meeting. At the end, he told me, "You were brilliant. Too bad nobody followed you." As product leaders, the roadmap meeting is our theater. We need to deliver great performances that inspire our employees to build amazing, winning products. Two key lessons I've learned from thousands of meetings are how to use "altitude" to keep employees with you across context shifts and how to overcome the *Penguin Effect* to get your team to test new ideas that week.

What Altitude Are You At?

I've had teams accuse me of changing my mind every week. The reality is that I'm often changing altitude, zooming in and out between execution, strategy, and vision, and failing to bring the team with me. If we tell the team what altitude we're at, they can follow better.

Every decision happens at a specific altitude:

- 5,000 feet—ground-level details: features, UX, daily execution
- 25,000 feet—strategic execution: How do the pieces fit together?
- 50,000 feet—big-picture vision: Does this align with our mission?

You know where you are, so the trick is bringing the team along for the ride. When we say, "We're going up to 50,000 feet," the team knows the conversation might shift dramatically. But that doesn't mean we're abandoning the 5,000-feet plan already in motion.

The Penguin Effect

Great leaders push their teams out of their comfort zones. Often, teams want to optimize for execution efficiency over innovation and growth.

At Zynga, we had created such a finely tuned jet engine that teams would find themselves in one of two states, neither of which was conducive to trying anything new:

1. They were crushing it—trying to keep up with massive growth, keeping servers up, content rolling. They were too busy succeeding to try anything new.
2. Or they were in *code red*—nothing was working, and they couldn't possibly afford to try something new and unproven. Doing so would be too risky an investment when the team was already struggling.

Cadir Lee, my brilliant CTO and past cofounder, called this the Penguin Effect—you couldn't get any of the teams to walk off their cliffs. But the minute any team tried something new and it worked, you didn't have to tell anyone else to try it. It went Zynga-wide within an hour.

I see the Penguin Effect in myself. The longer I go without shipping product, the more pressure I feel to ship. The more pressure I feel to ship, the more pragmatic and incremental I get.

Here's a harsh reality: Everything costs twice as much as you think and takes twice as long. That's my **Rule of Four.** If you apply this rule to building mouse nuts features—small, incremental changes—you've just wasted the whole quarter on something that won't move the needle.

Sometimes you need to take a different approach. How do you create space for real innovation when everything in your system is pushing toward incrementalism?

The answer is **Bold Beats.**

TL;DR

Your roadmap is the operating system for your whole company. It's where you translate instincts into buildable features with clear expected outcomes priced in engineering days. Roadmapping creates cycles of accountability that keep you and your team intellectually honest: You can see whether your assumptions were good or bad and whether you delivered. Used correctly, roadmaps connected to your OKRs can help you scale beyond yourself by getting people to do the right thing when you're not in the room.

6

Bold Beats

A Bold Beat is an attempt to deliver a high-impact innovation in your product that, when successful, creates a positive disruption in your user experience. Bold Beats should be quick to build, test, and kill, and have isolated impact on your product. When successful, Bold Beats should lead to major changes in your metrics. Think of them as controlled adrenaline shots for your product roadmap.

Most product teams fall into one of two traps: They either tinker endlessly with incremental mouse nuts features that no one cares about, or they get sucked into massive *boil the ocean* projects that take years and never launch. Both approaches kill momentum. The real magic happens in the middle—where you execute fast, bold moves that change the game.

We developed Bold Beats at Zynga out of sheer necessity. In a world where attention is fleeting, we had to constantly reinvent how we engaged millions of players. I encouraged the team to imagine they were summiting Everest with just one oxygen bottle. We shipped fast, took risks, and let the market tell us what worked.

Our bar for a successful Bold Beat was high: At least 25 percent of users had to try it, and it needed to move a key metric by 10 percent or more. But the real win was when a Bold Beat lit up social media, energizing our community and proving we'd struck gold.

Bold Beats don't just improve products; they transform mindsets. Teams shift from "What's the safest thing we can ship?" to "What's the boldest thing we can pull off in the next 30 days?" It's a shift from cautious planning to relentless execution.

Bold Beats provide a huge competitive advantage for any consumer product (and some enterprise products, too). While everyone else is stuck in analysis paralysis, you're shipping real, high-stakes innovations and getting immediate feedback. That's how you stay ahead.

A successful Bold Beat can add a new dimension to how your customer views your product and imagines where it will go in the future. And Bold Beats can be used in products far beyond games or software.

In FarmVille, an engineer spent half a day building "animals that move." While functionally this entailed just a few alternating GIF pictures, our players went nuts. On social platforms, they started to imagine many new dimensions to the game, such as horse breeding. This showed our game teams that they had many unexplored new zones in which to develop content experiences. It was a huge unlock for both game developers and players. After this, we also started to realize we could take any of the "fake" parts of the game that previously were just static art and turn them into new interactive game paths. "Buildings that matter" was a riff on this idea that enabled endless new content zones, as we could turn every building into a new game experience. A few examples of this included the barn, which increased item storage; an artisan workshop with crafting; and a market stall / order board, which enabled players to trade crops and crafted goods with other players.

The concept of Bold Beats extends beyond games, to any consumer product, even media. A great example in TV was the "hatch" that appeared in the first season of *Lost* and served as a cliffhanger for the next season. It blew my mind that this remote island had modern infrastructure. I think wondering what story lay beneath the hatch added a massive dimension for all of us; it kept 16 million viewers hooked for another five seasons. (As an aside, the creator of *Lost*, J. J. Abrams, is a committed gamer who loves puzzles and game loops, and you can see that in the way each episode answered one question while introducing two new ones.)

Of course, this approach isn't without risk—that's the point. Bold Beats force teams to answer one crucial question: How can we deliver something transformative this quarter, not next year? Constraints fuel creativity, and when you commit to bold, fast execution, you don't just ship features; you reshape the way people see your product and what's possible next.

> A *Bold Beat* is an attempt to deliver a positive disruption in the consumer experience. Bold Beats should be quick to build, test, and kill, and should have isolated impact on the overall product. A successful Bold Beat generates what Bing called an OMFG moment: It lights up social media channels. At Zynga, our bar for success was 25 percent of users tried it, and it moved a key metric by 10 percent or more.

Bold Beats at Zynga: CityVille, Mafia Wars, and FarmVille

We created a Bold Beat in CityVille when we added a single image of a construction crane on the far side of the river. There was no announcement, no explanation—just a mystery. The community buzzed with speculation on socials, and we monitored every comment, theory, and engagement spike. When we finally let players build a bridge over the river, unlocking beachfront properties, hotels, and a whole new world, it wasn't just an expansion. It was a *moment* that transformed how players saw the possibilities within their cities. Daily engagement and revenue skyrocketed. We followed the same playbook in FarmVille 2: We added a river as a Bold Beat. Suddenly, users could catch fish. This small, intuitive addition changed the entire trajectory of the game.

Because Bold Beats are meant to be isolated, temporary features, they enable fast validation of big ideas. In Mafia Wars, we built a system to track revenue and engagement metrics in five-minute intervals after every launch. When we introduced mystery crates—randomized reward boxes where players spend a fixed amount for a chance at items worth more than they paid—we sat together watching in real time as revenue doubled, then tripled, then quadrupled. The team was on fire. That's what happens when you nail a Bold Beat: You don't have to wait months to know whether it worked. You feel it instantly. Metrics pop. The team's energy shifts. Launching a successful Bold Beat always reminded me of Christmas morning. We knew it was going to be great before it even launched, and the anticipation never disappointed.

Bold Beats are more powerful when they're validated quickly with fast and frequent feedback loops. Used wrong, Bold Beats just become a fake name for new features. The power of Bold Beats lies in the fact that we give our teams permission to build a big feature in a "wrong," hacky way right now to see whether it even works. If the feature gets heat, then we can take the time to build it "right." A great example was "crafting cottages" in FarmVille—buildings where players combined collected ingredients into higher-value goods. We got a hacky version out in a month, which enabled our players to combine crops into new, useful items. They loved it and wanted more. We then took six months (a lifetime back then) to build this as a core part of the game system and long-term progression. This feature remained the biggest engagement driver for the life of the game. It was so successful that it became the basis for FarmVille 2.

Case Study: The Lonely Cow

Soon after launching FarmVille, we realized we needed a better way to bring players back to the game. Facebook had just opened their newsfeed to game content and encouraged us to send them anything cool we could generate. We realized we needed to create *social breadcrumbs* that would remind players to check back in on their farms.

One late night, FarmVille PM Jon Tien had the idea to create an emotional trigger in the Facebook feed. The Lonely Cow was born—a sad, lost cow that wandered off a friend's farm. The message? "A lonely cow wandered off [*friend's name*]'s farm. Can you help it get home?"

It was simple, it was silly, and it worked. The day after we launched the Lonely Cow, FarmVille saw one million more players. The mechanic created a perfect loop: It drove players back to the game, triggered newsfeed activity, and brought even more players in.

The Lonely Cow was a new growth paradigm. It turned Facebook's feed into an extension of the game. It worked so well we repeated it across every title: Treasure Island got the Wandering Sloth, and Vampire Wars got a Lonely Vampire Cow.

Successful Bold Beats Become Your Golden Mechanics

Part of my original thesis and innovation roadmap for Zynga was that we could bring every proven mechanic from hardcore games to mass-market players by making them accessible and social. (As a reminder, mechanics are game interactions that players want to repeat.) Eventually, we also developed new social mechanics, where interacting with other players was a necessary part of completing a core loop, such as turn-based play or gifting.

We first tested all of these as Bold Beats, and the most successful became what we called *Golden Mechanics*. These were proven features that became core building blocks for every future game. This was so crucial to our growth that we added a company value to remind us all that this was our job: "Innovate on social game mechanics."

Every product organization can build its own playbook of proven mechanics it can and should reuse. And Bold Beats are the best way to quickly build your playbook by testing your big ideas right now.

I want to call out two Golden Mechanics that first started out as Bold Beats: expansion packs and crafting. I believe they are still available as great areas to innovate on in gaming and in all consumer applications.

Expansion Packs

We learned from Bing that expansion packs were a huge ongoing business for traditional game franchises, and we decided to test the concept across our games. We soon found it was also a successful mechanic for casual social games. In Mafia Wars, we tested expansion cities.

Instead of just picking the next city based on gut feel, we added a "travel menu" with a list of potential cities and monitored exactly which ones players clicked.

My hometown Chicago won by a landslide, which might seem obvious given its gangster history, but the data confirmed our instincts. When we launched the expansion, players went crazy, because it was exactly what they said they wanted; we delivered the city they chose and a whole new Prohibition-era world of missions, weapons, and items to collect.

The moment we turned on Chicago, engagement exploded. Players who had never spent money before started buying virtual goods. These expansion packs, which started as a single Bold Beat, became quarterly events, which unlocked exciting new content and game features. Expansions became a Golden Mechanic we could repeat again and again.

And we did. Bangkok, Baghdad—each city was like a brand-new game, with its own economy and gameplay. Players could dive in, unlock new experiences, and bring treasures back to the main city. Every quarter, we treated expansions like launching a new season of a show or dropping an expansion pack for *The Sims*. It worked. Engagement exploded, retention soared, and revenue was through the roof. That's the power of a Golden Mechanic. We eventually used this same pattern to drive massive growth in FarmVille 2 and in FarmVille Country Escape through expansion farms.

Crafting

Crafting was the basis for massive hits such as *World of Warcraft*, but had yet to make it to casual gaming. *Minecraft* was released around the same time in 2009, but didn't become mass-market until a few years later. Crafting, which has been around for hundreds of years—from quilting to woodworking—provides the satisfaction of making something from scratch. In games, it's the same loop: gather, combine, create.

We first offered crafting in FarmVille as a Bold Beat, and it generated huge engagement—so then we took six months to build it correctly into what we call "Crafting Cottages." It became the most enduring engagement-driving feature in the game, and it formed the foundation for the design of FarmVille 2. Crafting is what made FarmVille and social games popular with Martha Stewart and her audience. In fact, years later, people speculated that social games caused the death of soap operas.

Gaming Golden Mechanics

When you find a Bold Beat that truly resonates, you can turn it into a repeatable pattern that drives sustainable engagement. It becomes a proven mechanic you can build on again and again.

Games aren't just play; they're master classes in behavioral design. We spent a lot of time in this chapter going deep on game mechanics, which are the perfect teaching tool to learn Bold Beats and represent the most fertile zones to test new ideas in any product.

Below, I've listed the most important mechanics that we implemented and innovated on at Zynga. Most are from consumer social games, but they can be applied to any user-facing product, whether it's consumer or enterprise. If you test these mechanics in your product, you'll probably see engagement lifts.

Traditional Game Mechanics

These have been proven over the past 40 years and always work, even to a fault. Beware the power of these. Users love earning and unlocking things so much that you may even generate false positive signal from these mechanics.

- **Coins, Badges, Levels, XP**—Rewards that drive engagement with core loops (Duolingo).
- **Leaderboards**—Rankings that show how you stack up against friends and everyone else (Zynga Poker, Strava).
- **Second Currencies**—Offer multiple currencies: one that is tight and valuable, usually representing real money, and one that allows (or many that allow) you to experiment with inflationary rewards. The more value you build into each currency, the more the currencies drive engagement in your core loop, revenue, and virality (Roblox—Robux vs. in-game currencies).
- **Collection**—The urge to gather a complete set of something (Pokémon Go).
- **Energy Boosts**—Gate play time on energy, so players pay or wait to keep playing (Candy Crush Saga).
- **Expansion Packs**—Extra levels, worlds, or features that keep players hooked after they've beaten the base game (Mafia Wars, *The Sims*).
- **Content Unlocks**—New levels, characters, or features revealed as you progress (Angry Birds, *Fortnite*).
- **Clans**—Experiences designed for teams to engage and compete in cooperative play (Clash of Clans, Peloton tags).
- **Quests**—Sending players on specific journeys can teach gameplay,

drive trial of new features, and introduce tons of easy added content (*Grand Theft Auto* missions).

- **Streaks and Drip Rewards**—Slow, steady dopamine hits (Wordle, Snapchat Streaks, Apple Watch Rings, credit card points, Duolingo).
- **Easter Eggs**—Hidden surprises that reward curiosity and spark sharing (FarmVille's secret holiday eggs that unlock rare decor, Spotify Wrapped, Tesla Fart Mode, Marvel end credits).
- **Loot Drops**—Random rewards found through gameplay (Starbucks app surprise bonus stars, cereal box prizes).
- **Mystery Crates**—Randomized reward boxes where players spend a fixed amount for a chance at items worth more than they paid (Happy Meal toys).
- **Limited Edition and Rare Items**—Collectible and time-limited must-haves (Disney Vault movies, Supreme drops, McDonald's McRib).
- **Crafting**—Collect resources and combine them into something more valuable. Traditionally found only in massively multiplayer online games (MMOs), but work for any user if made accessible and drive engagement with your core loop (*Minecraft*).
- **Gambling**—People will gamble on anything, and the more you vary the rewards, the more heat you create (Zynga Poker weekly tournaments—80 percent of players won, and some won a lot!).

Social Game Mechanics

These are game mechanics and conceptual frameworks that we innovated on and that drove so much heat they became core to our playbook and still apply today.

- **Gifting**—This was the first social mechanic we proved in Zynga Poker. Allowing users to send gifts of real value to friends is powerful to drive social engagement.
- **Make the Virtual Real**—This is how adults give themselves permission to play. Attach real-world connections such as money or social status and people care far more (real money gambling in Vegas, crypto).
- **Comms Channels**—Develop multiple ways for your users to communicate with one another. Similar to currencies, maintain one that is pristine and valuable so users know it's worth checking—for messages from friends, for example.

- **The Mormon Church**—Players always achieve Good, Better, Best. No bad outcomes.
- **Hard Gates**—Locking content or play based on a social interaction or payment (FarmVille required achieving neighbor counts to unlock more land plots).
- **Buildings That Matter**—Make every object interactive and fun, and make it lead to more depth in your experience (barns in FarmVille).
- **Crew Mechanic**—Hire your friends, and combine effort and resources to build shared assets (CityVille).
- **Status and Decoration**—Visible markers of achievement (Twitter/X checkmarks, Yelp Elite).
- **Community**—Offering ways for your users to meet and gather always over-delivers on engagement. People are lonely and always searching for more new friends. Many communities outlast their original services.
- **Neighbors**—Showing friends who are also playing on the game board to reward social interaction (adding neighbors to unlock content in FarmVille).
- **Slow Boat to China and Pay to Move Faster**—Always offer a way to grind to get to the reward. My rule of thumb is that for every 1 player who pays, there are 19 users who will jump through many hoops not to pay. Make those hoops fun for users so that they are valuable for you.
- **Turn-Based Play**—This is actually the best, organic hard gated play ever. In Words With Friends and Draw Something, players had to interact with a friend in order to take another turn. This created a social obligation to play.
- **Locked Content on the Game Board**—Seeing it all the time drives awareness and interest without any need for marketing. (We did this for the first FarmVille 2 expansion pack and sold $19 million worth of early access keys in two days.)

Bold Beats in the Wild

Bold Beats—and, at times, gamification—have been used outside of games to generate successful consumer products and marketing campaigns.

This gets back to the core philosophy behind life at the speed of play, which is that we can make any experience fun. In fact, the more boring the product category, the more opportunity we have to surprise and delight our users. Adding some gambling and mystery crates—or maybe even leaderboards—to your company's compliance training might be an OMFG moment!

Case Study: The American Express "Fun Card"

Bold Beats can work in any kind of product or service. I learned this first-hand with American Express.

I first met Ken Chenault, the company's then-CEO, when he was on one of these dog and pony tours of Silicon Valley. Zynga was considered a hot new game company, and so people wanted to come and shake our hands. (In every era of Silicon Valley, there are always hot new companies—today it's OpenAI—that every CEO and celebrity wants to visit to get the vibe.)

I overprepared and pitched Ken on what I would do if I were king for a day: create the next Fun Card. "If I ran Amex," I told him, "I'd focus on fun, not rewards. I'd launch the Fun Card. Stop competing on who gives the most rewards and make every transaction a game."

He loved the idea. His head of digital, Dan Schulman, worked with our team to create a gamified prepaid debit card. (Dan went on to run PayPal.) Every time you used the card to buy something, you got a notification that you won something in one of our games. The results were massive: Amex bought $70 million of Zynga virtual credits and ultimately gave away $300 million in consumer value. This drove 20,000 sign-ups for a new millennial-focused debit card and became Amex's most successful digital campaign at the time. More important, younger consumers didn't just use the cards; they started seeing them as fun.

Ken invited me back to give a talk to Amex's senior team and board about rapid innovation. I kicked off by saying, "I'm shocked you have me here, because I've been blackballed from your cards since I was 23." (I'd been blacklisted for 25 years over a $923 unpaid bill from my first job at Lazard. None of my companies could get an Amex card, which is a problem when you're ordering hundreds of data servers.)

In that talk, I looked at Ken and said, "Even though you're a much bigger company, I'm guessing that you have the same weekly frustration I do—you've got these managers that either are pursuing mouse nuts, little ideas that aren't gonna change anything but they're very measurable, or they're working on massive 'boil the ocean' projects that go on for so long people forget what the purpose was." Ken nodded his head furiously and said that's exactly what his life was like. I said, "I've got an answer for you: Bold Beats."

The gamified card was a perfect Bold Beat for American Express. We got it out fast and kept it isolated and focused, and it was massively successful, proving that Bold Beats can work anywhere.

By the end of my talk, Ken announced that I'd finally been approved for an Amex Black Card. Better yet, our pitch had turned a check-the-box meeting into a successful marketing play.

Case Study: In-Store Marketing with 7-Eleven

Game mechanics aren't just for games. In 2010 we partnered with 7-Eleven for its annual spring marketing campaign. The company spent $7 million on a billboard promotion and offered Zynga branded items on 30 different products, such as Slurpees and Big Gulp cups, in the 7,000 7-Eleven stores. For a month, the company's private-label water became "FarmVille Fresh Water," complete with codes under the cap for rare in-game items. And the stores rebranded their coffee as "Mafia Wars Coffee," which gave players similar in-game perks.

The mechanics weren't revolutionary. McDonald's had perfected the formula with Monopoly scratch-offs. But our twist was virtual goods. Unlike physical prizes, virtual goods have zero cost of goods sold. We could pack in way more value for customers, giving 7-Eleven the ability to sell products at the same price while paying us just pennies for the codes. The result? Their highest-ever spike in same-store sales.

During the six-week campaign, 7-Eleven's private-label products dominated the company's sales. It wasn't just a win; it was a roadmap for what retail could become. This remains a massive untapped opportunity to this day.

Case Study: Duolingo—Game Mechanics as a Growth Strategy

We've all seen a friend, a kid, or maybe ourselves become addicted to their Snapchat streaks (which reward consecutive days of play and reset to zero if you miss one). At Zynga, we refined streaks into a Golden Mechanic we called Drip Rewards: slow, steady dopamine hits that reward daily player engagement.

Bing introduced Drip Rewards to Duolingo in 2014 as an adviser, and the company integrated these and other game mechanics to make learning languages fun and a major mass-market category. As of October 2025, Duolingo served 135 million users and had a market cap of $14 billion. Duolingo is the best example of a consumer service effectively using game mechanics to turn a chore into a habit, to grow with almost no paid user acquisition or advertising, and to massively increase retention.

Leaderboards, leagues, badges, and streaks made language learning into a social competition that hooked people through status, progress, and shame. (Some users complained on social media that Duolingo was too aggressive in shaming them for losing their streak, but the shame worked.) These Golden Mechanics shot user retention from 12 percent to 55 percent.

In 2017, Duolingo shipped a great Bold Beat called Stories. These were snackable, character-driven narratives, such as "The Haircut" or "A Date," that made learning fun and sticky through emotional context. Learning vocabulary became more than just repetition. Users wanted to follow characters through their stories and development. While Stories were initially web-only, the mobile rollout drove a huge spike in engagement.

If streaks turned casual language learners into competitive players, then Stories turned them into fans. These two Bold Beats, and many more they've launched, have made Duolingo one of the most widely used and successful consumer applications.

TL;DR

Bold Beats isn't just a product strategy; it's a way to increase your team's ability to innovate. It's about dreaming of the biggest moonshot ideas you can imagine with your team and then challenging yourselves to turn those on in an isolated way for your users as fast as you can within your product cycle.

Bold Beats challenge us to prove that there's heat before we spend more time and engineering days on a product that won't deliver and a future that will never come. Bold Beats enabled me to scale Zynga faster, and they're still in place today in Zynga's quarterly OKRs. Done correctly, Bold Beats are one of the best tools founders and product makers have to get to what's right this week or this month.

7

Fuck Scale

Your number one job as a founder and CEO is to be right.

When I learned how to fly planes, I was taught that in an emergency such as the loss of an engine, the pilot needs to *aviate* (fly the plane safely), *navigate* (head to the nearest landing spot), and then *communicate* only after the landing is assured. This applies perfectly to running a product company. We need to first get our product 100 percent right before we worry about anything else, such as team, investors, and PR.

To build an Internet Treasure, you need to build something people love. Scale isn't the objective; quality is. So, to *aviate* means to consistently deliver a product people love.

It's always bothered me that McDonald's brags on its signs, "Over 99 Billion Served." That doesn't make anyone want to eat there. I'd rather see "The Best Burger You'll Ever Eat." Nobody cares that your company made and served a million hamburgers today. People just care that their burger is perfect. That's why, in California, we all want to go to In-N-Out Burger over McDonald's. And it's why you'll probably never see In-N-Out across the country. They care about quality, not scale.

And of course, the day I was writing this chapter, I saw Garry Tan, the talented head of Y Combinator, post on X . . .

The question isn't how to serve more customers—it's how to serve them right. Nail the product, and the rest will follow.

Japan values quality over scale. Anyone who's been there has experienced this ethos of quality for the sake of quality. I visited a bar that would allow in only seven people at a time, because that's how many the bartender could serve well. He spent 5 to 10 minutes preparing each drink and served it with handcrafted ice cubes. This ethos is the reason Japan has had a big impact on Steve Jobs and every other product maker who's spent time there.

So then, what is the role of scale and why should we care about it?

If we're successful with our "right product," we likely will have to grow our teams and inevitably reach a point where we can no longer manage every detail. So we will have to delegate. Therefore, I define scaling as getting people to do the right thing when you're not in the room. Recently, I surprised classrooms of MBAs at Stanford and Harvard when I argued that this is the only point of management.

This chapter covers three core principles of scaling:

1. Being right.
2. Getting people to do "right" when you're not in the room.
3. Not being distracted by the forces that destroy "right."

Core Principle 1: Being Right

Jeff Bezos said the number one criteria at Amazon for a product leader was that they had to be right. If product leaders couldn't get Amazon's product right—meaning offering the lowest price and most selection—nothing else mattered. In fact, one of Amazon's Leadership Principles says that leaders "are right a lot." BTW, I like "right a lot" better than "right." Nobody will ever hit 100 percent, and we want risk takers.

Being right comes from staying close to the metal. It means getting into the details about your users, your product, and the teams building it. But being right isn't just about execution; it's about taste. Steve Jobs nailed this with the iPod. Plenty of MP3 players had similar specs. But Jobs insisted on something simple and elegant: a scroll wheel that felt intuitive and packaging that felt premium.

Humanize Before You Productize

As we discussed in chapter 3 with Minimum Idea State, the fastest way to test and learn is often to build your experience by hand first, with an approach that I like to call "humanize before you productize." Airbnb has lived this ethos, from its founding to how it continues to invent new products today.

Recently, Joe Gebbia, cofounder of Airbnb, shared the long-form version of the company's origin story over dinner. His path to product market fit sounds amazingly simple and obvious in retrospect. It's both painful and inspiring to think that any of us could have done this. Joe and his cofounders were still in their Y Combinator batch, and Airbedandbreakfast.com had been flatlining for six months. They met with Y Combinator creator Paul Graham, who asked where they had any customers. They told him they had 30 hosts in New York City, and he replied, "Then why are you here [in Mountain View]?" In response, Joe visited every host, offering free professional photographs. People were shocked that the website founder showed up at their apartment and were happy to hang out and chat about the service and other options they were using. He convinced them all to try lower prices to start generating reviews and more renters. *And* the hosts all shared their lists of complaints and improvements. One even pulled out a notebook. Many complained that the calendaring sucked. When Joe emailed his hosts two days later with a link to the improved calendar, they all became superfans. Then, something magical happened, which created the Airbnb we all know today. Joe and Brian saw the number of their hosts double the next week to 60 and keep doubling after. A few months later, they saw listings pop up in Iceland and Turkey. That's when they realized that what they were selling was a trusted community, and their community was spreading this service with every interaction.

Even as it's grown, Airbnb has continued taking an anti-scale approach. Around 2016, Brian Chesky invited me to see the final version of Airbnb Experiences, which he'd been working on for years with a handpicked team. They were creating epic experiences by hand and looking for OMFG moments from small groups they tested these on. They'd curated experiences with top chefs such as Michael Mina and famous actors including Sarah Jessica Parker.

I told Brian that Experiences was a cool idea, but it wasn't what anyone thought of when planning a typical trip. If you're going surfing in Maui, you may want great rental boards waiting and even a local coach to paddle out with you. But you're not looking for Michael Mina to cook dinner. In fact, having to fawn over a famous chef can feel like a chore when you really just want a steak and tequila. The team started calling this "The Pincus Problem." I started meeting with them weekly and helped with a redesign for three months, which culminated in the launch of Experiences, which was a first-base hit for Airbnb, peaking at $175 million in gross bookings in 2019. Brian hasn't given up on his underlying instinct that people want more than just an easy room booking, and in May 2025, he invited me to his Summer Release, where he relaunched Experiences along with a suite of new features.

Great product makers stay close to the metal; they know taste lives in the details. We build and refine our taste by noticing every product we use. I'm constantly noticing my own experiences with other products. When I use Uber and get frustrated by estimated arrival times that feel dishonest, I know other users feel the same way. That's why I told Uber CEO Dara Khosrowshahi that the company should test adding a timer and reporting the percentage of on-time pickups that day. These might look bad initially, but could rebuild trust in the brand and create a better product experience.

Focus Group of One

Like building your instincts about taste, being right about audience traction comes from constant testing. Often, we can find our best signal with a *focus group of one*, where we can just tell that individual perfectly represents our target audience and their sensibilities.

Every time I'm with my niece and four nephews, and often with my teen daughters, I ask them about their favorite games and, even better, their reactions to my ideas. Recently, I rode a gondola with two 13-year-old boys. I asked them, "What if, in *Grand Theft Auto*, after you beat up businessmen and take their money, you come back an hour later and see neighborhood watch posters with your photo?" They said that would be sick. That was more impactful in validating my product vision than any focus group or heat test I could ever construct. I shared a user story, and it deeply resonated for them. It had adrenaline. They clearly wanted this product and started imagining what else might happen beyond it.

That's the reason a focus group of one often matters more than any other data or feedback. Our job as consumer product makers is to keep telling stories about the products we dream of making, looking for the heat. Your focus group of one should start with yourself, tuning the product to your own taste. In games, I hate the fake parts—I want to believe it's a real world. That core instinct has driven me to make the virtual real. It's what led to FarmVille. But we have to rapidly and accurately validate that our taste actually will translate to millions.

Emmett Shear figured this out with Twitch. He succeeded only when building something he personally wanted to use. With Justin.tv, he and his team thought they were building for viewers, but they actually built a great product for producers, because that's what they themselves were. When *StarCraft II* came out, though, Emmett found himself watching gaming content constantly. He had an intuition: "I don't think I'm that weird. If I really love this, there's probably a bunch of other people who love it as well." He did the market analysis afterward, but the core instinct came from trusting his own taste as a user.

The best product makers can talk to a handful of people or see a few key pieces of data and know their product is right—or, more often, that it's wrong.

Core Principle 2: Getting People to Do the Right Thing When You're Not in the Room

Your greatest leverage as a founder and CEO is to build a culture in which your team consistently makes the right decisions when you're not in the room. First, I have to say that I die a little inside when I see myself write "culture." I have always had an allergic reaction to this word and all the books and talks telling us how we need to focus on this. It reminds me of the first rule of Fight Club: Don't talk about Fight Club. Culture is an output of all you do, not a conscious input. It develops organically as a result of many small and sometimes big actions as you build your product and company. But how do you develop a culture without turning it into a buzzword?

The Moral Contract

Your best chance to get people in your company to do the right thing when you're not in the room is to create a *moral contract* when you *are* in the room with them. That starts when they join your company.

What is a moral contract? It's an unwavering bond with every member of your team that, if they deliver, you will recognize and reward their contribution, often with a bigger role and more compensation. But it goes beyond this transactional definition. It's based on the premise that we are all going to war together. You've got their backs. You're on a longer arc together than this one product or company. Cadir Lee, Scott Dale, and 33 others from my previous companies joined me at Zynga. As a leader, you have to back up your words with your actions, demonstrating that this relationship matters more than any paper it's written on. Your teams can and will hold you accountable to this contract, and you won't always live up to their expectations. But you may be the first founder they've worked with who has offered this up and tried to live by it. I also want to be clear that this moral contract is not a blanket promise that you will take care of everyone who comes to work at your company. As Reid and others have said, this isn't a family. The moral contract is not unconditional. It's saying that if they give everything they have and help deliver amazing outcomes for our customers and our company, they don't have to worry that their work will go unnoticed and unappreciated—which is the norm in most companies.

Marcus Segal, an early employee who left and then came back to Zynga as Chief Operating Officer in 2015, called the passion I tried to instill in people—especially about product—"passing my vampire blood." You give people your vampire blood when you sell them so hard on your vision and mission that they bleed for it. At Zynga, we used to say that they "bled Zynga red" (which was our brand color). When you're leading a smaller team, you have the time to do this with every new employee, and even when your team is bigger, doing so is still worth your time (perhaps even more so).

Micromanagement Is Beautiful

You should micromanage as long as you can. At Zynga, I ran a daily stand-up call and managed every engineer through a Google spreadsheet with every name and deliverable, until we got to 50 people and the daily meeting became two hours long. The call created a daily accountability that nobody could hide from and a sense that we were working on a 24-hour execution cadence.

Beyond 50 people, I was forced to manage through weekly product meetings. I tried to have as many team members as possible in the room. I found that I could still manage the company, up to 500 people, through force of will, which resulted in weekly meetings with 14 different product teams.

My point is this: The more often you're in the room, the better the chances that the team will make the right product decisions—and you can kill a few birds with one meeting. Pass on your vampire blood: Teach your team how you make the right decisions and inspire team members by recognizing their great ideas.

Never Do One-on-Ones

If we're maximizing our time spent on great products and leveling up teams, we can't afford to do one-on-ones. First, these meetings generate more politics, because once there's a precedent, everyone will want their one-on-one meeting. This rewards your extraverts and penalizes the quiet, humble contributors. And these meetings eat up your calendar. I promise your life will improve once you eliminate them! I learned this indirectly from Jeff Bezos via Bing Gordon, and once I tried it, I was a believer.

While you will inevitably still have to meet one-on-one with people at times over compensation or personal issues, the bulk of your operating time should be spent meeting with your leaders and their teams. You can touch many more people and give them exposure, too. And if anyone ever corners you and starts complaining about other team members, the best thing to do is to call the other team member or members into the room, tell them what the first person said, and leave. Your message is that you won't tolerate politics, and you will save massive time, drama, and cycles. This works similarly with my kids (most of the time).

Keep Your Hands on the Wheel

Some decisions only you can and should make as a product founder. Jason Citron and his cofounder, Stan Vishnevskiy, realized this at Discord. At the time, they were outsourcing the most important decisions—features touching end users—to their most junior product managers.

Jason shared with me that he and Stan eventually realized what a big mistake they had made in their effort to scale the company and organization, so they grabbed the wheel and mandated that they had to personally approve any feature before it went out. That's quality control. That's how you guarantee excellence, even when you're not in the room. Because even if a feature was built by someone else, it still has your fingerprints on it.

It's the same way great chefs run a kitchen. They have sous-chefs, but they still walk around tasting everything to make sure the quality is in every dish they're serving.

In this regard, Silicon Valley is coming full circle. For the past two decades, there's been an obsession with scale. A lot of that was driven by VCs and the professionalization of tech. Before Amazon and eBay, technology was the Wild West. Then, management consultants from Bain and Mc-Kinsey started coming in—people such as Meg Whitman at eBay. We all thought, *Wow, these people know how to manage at scale.*

And for a while, we all looked up to that. We wanted to be eBay and Amazon. If you were a start-up founder, the idea was to get someone from Amazon or eBay to teach you how to run a big company. When I was building Support.com, my VCs pushed us to recruit people who had managed at scale. That's the same argument my VCs used when they replaced me as CEO: *You've never run a large company before—why should we risk you learning?*

But here's what we all missed: We were over-indexing on the management scale principle, instead of the "be right" principle.

Now we're seeing a renaissance of product founders who are in the details: people like Brian Chesky at Airbnb and Jason Citron at Discord. Part of the reason is that Steve Jobs was so successful and Apple has taken over the world; but we've also grown up enough to be confident in ourselves. We're no longer held back by the old VC standard.

Creating Product Leaders

Your best chance to grow and maintain quality is by creating strong product leaders who can be right. These leaders can almost always only be developed organically inside the company. Even if you can recruit talented, experienced leaders, they will need to learn your product formula and processes before they can be effective and right.

Jon Tien, whom I mentioned earlier, created the most successful Bold Beat in FarmVille—the Lonely Cow—as a new hire from Boston Consulting Group. He did this even though no studio in the company was willing to take him on board because he had no product experience. Skaggs and FarmVille were desperate for people, so they had no choice, and they took Jon on as a junior (and initially, the only!) PM on FarmVille. He was so talented that he not only drove the most successful Bold Beats on FarmVille, but, within 18 months, he went on to become head of product for the whole company at age 26. Because he learned product management at Zynga, he knew only Zynga's way of doing things, which made him more powerful (and faster?) at driving Zynga's product way across the whole company. He went on to lead product, first by cofounding a company, and later at *The New York Times* and now the mental health company Calm.

Sim Singh joined Zynga Poker in early 2008 as a junior PM, and he was so talented that we gave him two battlefield promotions in the space of four months. He became the GM of Poker at 25 or 26. Spot promotions for amazing work can surprise and delight great contributors, too.

Ian Cinnamon was a brilliant kid who had grown up making his own mobile games. We hired him out of MIT and had to compete with Meta and Google to get him. I recruited him and told him I would personally manage his career at Zynga, because I saw so much potential in him. I put him inside Poker, but there was organ rejection—he was so entrepreneurial and self-directed that the Poker team tried to fire him from the company. (I ended up making him my tech assistant, a role I'll explain in the next section.) He worked personally for me on projects and went on to found several companies that I've invested in, most recently the hot space start-up Apex, which is manufacturing satellites and has raised hundreds of millions of dollars at a valuation of more than a billion dollars.

Tech Assistants

Ian is a great example of how powerful a tool tech assistants are—they're a way to clone yourself while growing new leaders.

The tech assistant concept was first implemented by Andy Grove at Intel and was later embraced by Bill Gates at Microsoft and Jeff Bezos at Amazon. The idea was to pick a high-potential contributor from your ranks and basically make them your shadow. One summer at the Allen & Company Sun Valley Conference, Bezos spent a few hours explaining the concept to me, and I adopted it at Zynga. TAs spend 12 to 24 months sitting in every meeting with their CEO, and then they dive deep into research and special projects. Today, nearly every Amazon C-staff member came from this role, including the current CEO Andy Jassy.

Over the years, I've had six TAs, most of whom have gone on to found their own companies. Zynga TAs have included Michael Chow—who has founded multiple game companies, including one acquired by Riot, and is now working on a gaming company, The Believer Company—and Dan Garon, who went on to work with me post-Zynga across several companies and is now founding his own start-ups.

Make Everyone a CEO

Early on at Support.com, I discovered something that would become part of my product management bible. When we grew to 35 people, I put a big 3M sticky sheet on the wall and wrote down everyone's name. "By the end of the week," I announced, "you need to write what you're CEO of, and it should be something that matters to everybody else. You'll own it, and everyone else will know to come to you with any questions about it."

One breakthrough came when our receptionist needed to buy a $500K PBX system. Instead of just approving it, I told her, "You're the CEO of our office. Research it and present it to me and the board." She did, and she went from answering phones to managing our entire phone system, eventually becoming an HR leader.

The lesson here is to push people to play the position above them—give them so much responsibility that it scares them. The best people will rise to that standard and love their jobs more. I don't believe in paying dues. There's no time!

This concept of making everyone a CEO became a core management philosophy, which I applied in new ways at every company. It became a central building block to grow Zynga. Bing added that for big leaders, you should hire and promote only people you would back in a new venture with your own money. He said that Bezos had applied this concept at Amazon, saying that if it were the 1500s, the person he would hire as a leader would have to be someone he would be comfortable handing a treasure chest to and assuming they would return in a year with a bigger fortune.

The problem with most companies is that people have narrowly defined jobs and require a lot of management and added layers. I've always found that the best leaders are homegrown. Take junior people with willpower and firepower and make them the CEO of their jobs. Most people consider

Satya Nadella Microsoft's best leader ever, and he was a lifer at Microsoft, spending 22 years there before becoming CEO.

Making everyone a CEO enables you to scale a company and a product organization. In the early days at Support.com, our team of 35 was able to act like a team of 350.

Battlefield Promotions

Overpromoting people can be an awesome strategy to generate more leaders. Often, we find great young talents who are capable of more. Try giving them a bigger job; they will often surprise you. It sends a great message that your company is a meritocracy, while also creating more room for others to grow. I've found that when people are given huge responsibility, they care more and grow faster. Generally, this practice contributes to a culture in which everyone is playing the position above theirs with an expectation they will soon fill it if they deliver.

This practice also sends the message that we have no room for "rear admirals." Every leader is in the trenches executing on product, or they're out.

When you run out of people in the trenches to promote, you often have to hire outside leaders.

Hire Beneath the Current Position

Successful product companies will hit growth curves that quickly outpace their ability to manage. In this AI era, it's become common to see companies hit $100 million annual run rates within two years, and the current "cool" thing is to brag about how small your team is. While I love small and scrappy, the reality is that we often face moments when we need to hire outside leaders or else miss growth opportunities. The challenge—and often the big risk—is that because these people haven't organically grown up inside our unique company, they will come with outside beliefs and practices that could destroy the very formula that made us so successful.

How can we do this successfully? I failed many times at Zynga and Support.com with big outside hires. In fact, it became a running joke how many COOs Zynga ran through. I'm guessing we had at least seven or eight.

The way we got it right the most often was by hiring a leader into the position below the role we intended for them. That meant hiring studio heads first as GMs or VPs first as directors. The reason to put them lower in your stack is that they need to earn their stripes inside your product and your audience—and, most important, win the respect of your team.

When I recruited Mike Verdu, he was our first outside product leader. It was a coup to get him to leave EA, the top game company. But he had never run a live social game before, so how could I have handed him the wheel for one of our biggest games? Luckily, Mike came with incredible humility and chose to run Vampire Wars, one of our smallest games. He correctly realized he needed to create one successful feature himself, in order to learn our business and gain respect and credibility across the company. By doing so, he quickly felt like a Zynga core team member and not a transplant. He spoke our language with authority, and teams followed.

Unfortunately, we rarely found people with the same level of experience and humility as Mike, so we had to find new ways to grow successful leaders that could get to larger pools of potential strong people. At one point, I decided to hire a whole group of smart friends who had been strong CEOs of failed start-ups. None had any experience in games. I had most start off as GM of Poker because it was the hardest game to screw up. They learned social gaming, which was as much about running a live 24–7 service with its own P&L as it was about software development. They graduated from this line-of-fire training program to go run new studios. These successes led to one of my go-to hiring hacks, **Broken Résumés**, which I've applied at every stage of company creation and growth.

The Talent Myth and Broken Résumés

Everyone tells you that your first 10 to 20 employees have to be amazing. It's not wrong that people matter, but this rule creates an unrealistically high bar. This **talent myth** puts so much weight on the caliber of people needed that it becomes overwhelming or even debilitating to make those first hires. As a result, we often over-tune to résumés and past experience in looking for these unicorn first hires.

The reality is that most new product companies sound flaky before anything is built or proven. That's been the case for everything I've ever launched. What we really need are people who show passion for our product and who are available. Often, these people are around and interested because they lack better options.

One hack I use is to look for Broken Résumés—people who came close to achieving greatness but failed. These people are way more available and come in with hunger, humility, and curiosity.

At Zynga, I also hacked the "Founder" title to recruit and motivate early people. We called everyone who joined in 2007 "founding team members." And when I think of the standouts, they were people like JW, the 19-year-old prodigy who went on to run the YoVille studio, and Neil Souza, a PHP contractor right out of college without any work experience who became the engineering lead on Mafia Wars. We hired these two because they could code, but we soon discovered that they had many other hidden talents—and they loved games. They brought a taste and aesthetic and a little bit of a pirate culture. With FreeLoader, I hired government Lisp programmers who ended up being awesome. Without the experience of working at a big company, they attacked problems from first principles. That kind of homegrown talent led to an unexpected magic and uniqueness that you can't replicate with big-company hires.

Adopt a healthy aversion to people whose résumés show they're from the biggest and most admired companies. People who come with A-list experience often act as if they're doing you a favor, but they're not. They're not coming in ready to invent new lessons alongside you and your team. You don't want teachers; you want learners. You want to coinvent what is going to work.

Another pitfall of this talent myth is that it can cause us to waste too much time searching and interviewing. That's why at Zynga, I used hiring as my interview process. In our first year, I didn't have the time to waste finding the best people. Honestly, I'm terrible at interviewing anyway. So I figured, why not just hire anyone who knew Flash or PHP, and keep the people who were productive? That's where Neil came from. No recruiter or LinkedIn would have ever surfaced him.

> ### Easter Egg
>
> Look for Failed Versions
>
> It can feel impossible to recruit your initial team from any company, much less from your ideal target. When I was starting Zynga, I couldn't get anyone from a top game company like EA to consider us—at first. However, I've found there's always a deep well of talent behind failed attempts to build earlier versions of our products. In fact, there's a million new mobile apps each year that fail. I can guarantee you that the person who made the 250th WhatsApp replacement will be happy to hear from you. There's a good chance they share your passion about what you're doing and would've kept going if they didn't run out of resources. They probably got there with a nontraditional path (no Ivy League degree or big-tech experience), which means they're not getting tons of inbounds from recruiters harvesting LinkedIn profiles. I've always said the best résumé is a well-made product—which you have right in front of you. It's the app they just made.
>
> The bonus of a nontraditional background is that you're much more likely to build real diversity on your team and in your founding DNA early on. By diversity, I mean people with much different backgrounds than the people you've spent your career working with. I learned with Zynga how powerful it is to get outside our narrow gene pool and bring in people without college degrees, people who are self-taught. The most important thread at Zynga was that people were in love with social gaming. They were passionate about the mission and that trumped the résumé.

Challenge Assumptions, Not Judgment

One of my guiding principles in partnering with strong leaders is to challenge their assumptions but trust their judgment. If you trust someone has good judgment, then you can work with them—even when they're wrong. The key is to separate bad assumptions from bad judgment. You need to make sure they are chasing the same hill as you and are making the correct assumptions to take that hill.

Good teams challenge decisions by questioning the assumptions behind them, not the person making them. When you start to question someone's judgment, it feels personal to them—and it starts to bring into question whether they're in the right job.

If you can't trust their judgment, then you can't trust them to do the right thing when you're not in the room.

Reinvent the Corporate Stack

Great companies don't win just because they have the best talent. They win because they have more people who care. Your "corporate stack" is every touchpoint that gets people to care—mission, values, compensation, promotion. It's telling them how and where you're going to innovate and win. Ultimately, it ensures people will have the right point and argument when you're not in the room.

You should develop these values in the trenches, organically, as you prosecute your product. Every value is earned with blood and sweat as you take key hills. Values almost always have stories behind them. And these stories tell your team what winning looks like, inside the company and in the market.

At Zynga, we proudly displayed our values on the wall. The first value we put up was: "Make games you and your friends want to play." My hope was that it would make people raise their hands when we were making games and features they didn't get or like. And this worked, although it also bit me in the ass sometimes, like when I couldn't get anyone initially to work on FarmVille.

Later, we added what became the most important value, which was to "innovate on social game mechanics." It told everyone that this was their job: to invent fun new ways for people to play together, not better tank explosions. That was actually how we grew our company.

Core Principle 3: Don't Get Distracted—the Forces That Destroy Right

We tend to spend far more time planning for things going wrong than for everything going right. When things finally do go right, the problem we face is that we're not ready for what comes with explosive growth—all of these other forces that come in and distract us from our true mission and purpose, to build great products that people love. All of a sudden, we're elevated to some greater position in the world: The press wants to cover everything we do, and we're invited into rarefied circles of top entrepreneurs, CEOs, investors, and even celebrities. One time, will.i.am reached out to us to visit the Mafia Wars studio because he was playing the game. These exciting fun things happen, and we see different founders handle them in different ways. Unfortunately these forces rarely or never allow us to put more time into our product, which is our job, our growth engine, and, ironically, the reason the world cares about us. I want to talk about a whole bunch of forces that pull us away from our mission, which is to be right and to continue to engage with our teams, to be in the trenches so that we can deliver products that surprise and delight our customers.

Micromanage Your Own Time

As Zynga got bigger, Bing gave me a rule to keep me honest: My tech assistant and executive assistant would look at my calendar and grade whether I was spending at least 50 percent of my time on product. If I was ever less than 50 percent on product, I was "redlining." The only time Bing ever challenged this rule was during a weekend breakfast when he told me that if I didn't spend more than 30 percent of my time hiring big product leaders, we wouldn't be able to keep growing in six months.

As your company grows past 50 people (and even before that point), it becomes essential to step back and manage your priorities and focus on your time. Look at your calendar through this lens: How are you maximizing your impact on your product, teams, and company?

There are things on your schedule that are clearly good and impactful for your product, and then there are things that clearly have nothing to do with your product. You want to crowd out the things that don't have anything to do with your product.

What should be on your schedule? Roadmapping, product meetings, brainstorms, maybe recruiting for more product people and engineers. Any time with your product team, especially when you are small, is going to impact your weekly product delivery. What shouldn't be on your schedule? Talking to investors, going to conferences, talking to the media and lawyers. You want to cram those down into the least amount of time possible. Ideally 80 percent of your time is spent on things that impact product, and 20 percent is spent on things that don't.

Don't Be a Fake CEO

Bing and I had a mantra at Zynga: "Don't be a fake CEO." We both called it out any time I was stepping into danger zones that would push me further away from our players, products, and teams. My fake CEO signposts were activities such as talking on panels, hiring COOs, and doing magazine photo shoots. A perfect example is the one the board and I did for *Fortune* with a chicken sitting on my lap. (Reid was hugging a goose and Bing was shucking corn.)

We actually had mostly negative press about Zynga. So I'll admit that in 2010, when *Fast Company* asked me to be on the cover of the magazine's "most innovative companies" issue, I jumped at the chance. But I didn't want to have my "fur coat moment"—that moment in *American Gangster* when Frank Lucas appears on the front page of *The New York Times*, which leads to his ultimate downfall. My highest priority was whether a nurse in Indiana liked FarmVille and wanted to share it with her friends, not that I was in a magazine.

Photography by Gregg Segal

In fact, in 2008, I had hired Brew PR to keep us out of the press. At the time, we were making huge money from users paying for virtual goods, and we didn't want to share that with the industry. It was a trade secret. So we allowed others to perpetuate a false narrative about the company: that we made all our money from ads. It got to the point that *TechCrunch* even wrote a story called "Scamville," alleging that we made all our money from aggressive ads and offers, which wasn't true. It didn't matter. Our users didn't read *The New York Times*. They cared about what was on their friend's Facebook profile wall.

When you do get offered high-profile meetings and speaking opportunities, there's a right way to use them. For example, when Carol Bartz, who was then the CEO of Yahoo, asked me to speak in front of her senior team, instead of doing the expected talk show–style interview onstage, I presented on how Yahoo could become the world's biggest social gaming portal. I started with mock-ups of a new Yahoo home page, preloaded my slides, and turned what could have been a check-the-box meeting into a revival-style pitch for social gaming. It led to real results—a test integrating game requests into Yahoo email.

The same principle applied when Allen & Company included Zynga when touring top public market investors through Silicon Valley. The company told me I didn't need to prepare and that investors just wanted to meet me. Instead, I treated the meeting like an IPO road show. And I did the same when they came back a year later. I built credibility and relationships with investors long before we got to an IPO. If someone's giving you an hour with key decision-makers, overprepare and make it count.

Control Your Destiny: The Facebook Fight

As I shared in the Book of Life chapter, my dad instilled a deep sense in me that as founders we can never give up control, no matter what. This was put to the ultimate test when Facebook threatened to turn off all of Zynga's traffic if we didn't agree to be a captive company.

The irony was that I had spent years trying to convince the Facebook team that what we called user pay (today called in-app purchases) would be a huge business for them, and then they tried to kill us for proving it.

When Facebook first launched their app platform, they had no intention of hosting games. The app they envisioned was Causes, cocreated by their former COO Sean Parker. While it did well, it never became a top app. I quickly became an evangelist for social games inside Facebook, showing up weekly, talking to Zuck or anybody who would listen. I had one ally inside Facebook: Jared Morgenstern, their one revenue PM (who later went on to be COO of Raya). Jared loved the idea of user pay and was passionate about it becoming Facebook's biggest revenue generator.

By 2010, social gaming was Facebook's biggest user engagement driver outside of their News Feed. Zuck's original vision for a long tail of app developers never happened. Instead, Zynga's network of game apps dominated everything. We were the 10 biggest apps on the platform. Zynga grew to 20 percent of their page views and 10 percent of revenue.

In fact, we were too big and making too much money. I heard that people inside Facebook started talking about "the Zynga problem." They felt we were using more than our fair share of the platform and of their users' time and attention. They introduced Facebook Credits as a voluntary payment service, and like Apple Pay today, they charged developers 30 percent of revenues. (Ironically, later, Zuck publicly complained about Apple doing the same thing.) The program was "voluntary" because Facebook had promised their ecosystem would be open—they told the entire developer community to come use their APIs, no agreements needed, and the understanding was that they would never restrict access or charge. One promise they eventually broke, but whoever hosts the platform sets the rules and developers will go along with almost anything if they can get traffic. (Years later I heard Microsoft CEO Satya Nadella say that a platform should deliver 10 times the value in its ecosystem to its own market value. Also ironic to hear from Microsoft, which spent decades hoovering up any value it could at developers' expense, but it's never too late to find enlightenment).

We actually wanted to pay Facebook, hoping that would make them take games more seriously. Games have always been treated as second-class citizens on platforms. In the case of Facebook, we were constantly fighting to keep our persistent navigation on the home page and our players' access to their communications channels. Any Facebook product manager could wipe us out to promote a new internal feature or app like Events or Birthdays. It felt like riding on a five-story high unicycle. We kept adding more stories, but the platform could and would change on the whims of its various PMs, and all of us developers would have to scramble for survival (and many didn't make it). When they changed the platform, which moved and sometimes deleted our persistent navigation, it was a code red for us and the whole developer ecosystem as everyone's metrics nosedived. I wrongly assumed that if Facebook was generating big revenue from Zynga and others, they wouldn't allow this to happen anymore. I was just off on the timing. Zuck didn't care much yet about revenue. That came a decade later when they were public and needed to show quarterly revenue growth.

When we eventually did test their payment system, it was a disaster we could never have anticipated. It was so much worse than our own internal payment system. In our tests we found that 30 to 50 percent of transactions were denied. We believe this was because their system was set to more conservative filters on transaction verification. And given no developers were choosing to use it, we couldn't afford to be at such an extreme competitive disadvantage. Imagine today if Apple were forced to open its App Store payments and developers could choose to use PayPal or anyone else for a fee of 0.5 percent of revenue. Zero would choose Apple Pay. Only a platform monopoly can force that kind of fee.

Early in 2010, Zuck called me in for a meeting with him and Sheryl Sandberg to understand why we weren't using their new payments system. Zuck said if we adopted it, all the other developers would, too, because they were all following us. He had Sheryl explain it, saying that she had worked at the US Treasury Department. She talked about "free riding" and "paying your fair share of infrastructure costs."

"I'm happy for you to make money on this. Just require the whole ecosystem to do it, and we will, too." They kept saying I wasn't "leaning in" and they needed to see Zynga "lean in."

Facebook punished developers by giving them "moratoriums." They'd find a minor infraction, and on Friday at 5 p.m., they'd turn off our access to their APIs. Basically, they'd starve your app. Then, because this happened end of day on a Friday, the Facebook team would be gone. We'd freak out, try unsuccessfully to reach anyone, then suffer all weekend long. It was like one of those movies where they're trying to run you out of West Point on minor infractions. We felt that we were held to more scrutiny and to a far higher standard than any other developer.

In the spring of 2010, a Facebook exec read us the riot act. One Friday afternoon, he came over to the Zynga office and pulled a classic good cop, bad cop. It went something like: "I'm the only friend you have inside Facebook, and I'm the only thing stopping you from being totally shut off. People inside Facebook have had it with you and games. The one saving grace was for you to voluntarily sign up for credits so at least you'd be paying your fair share. We tried to be nice, and you guys aren't playing ball. So now we're giving you an agreement to sign. You have until Monday. If you don't sign it, we're going to have to shut off all your traffic."

The agreement would have rendered our games exclusive to Facebook, meaning we would never be an independent company. Not only would we be the only developer to commit to credits, but our games could never move beyond Facebook. It's hard to imagine today what would have happened if we had agreed to never publish our IP on mobile other than through the Facebook app.

This reminded me of the FreeLoader moment when David Wetherell told me I had to hire a new CEO and be exclusive to Lycos. These kinds of ultimatums represent existential threats and are our biggest tests as founders. It's where we find our deepest resolve. We didn't choose the easy, safe path. We're risk takers who stand for what we believe in, and that means never being captive to anyone.

I also remembered something Zuck had told me. I'd been going on walks with Zuck, and he'd been going on walks with Steve Jobs, and he shared with me what Steve said: "Never let any of your developers be more than 1 percent at most of your platform because they'll get too powerful."

I went back to the team and the board. "There's no way we're signing this. It's like selling our company without getting paid."

At no point did John Doerr or anyone on the board suggest we take the deal. They all said, "You might as well say no, because they're going to kill us anyway." We decided to take our chances, which meant fighting the very platform that we required for our lifeblood.

We didn't sign the agreement. Facebook started giving us more moratoriums.

About two weeks into what had become the Cuban Missile Crisis between our companies, two things happened: First, I was told a lawyer from Facebook accidentally emailed us with their internal conversations, which I'm pretty sure mentioned the "Zynga tax." Second, Reggie Davis, our general counsel, realized we'd never signed a single agreement with Facebook—not even an NDA. There was nothing stopping us from publishing their agreement, although we never did.

That would have seriously harmed Facebook's credibility with the developer community and their bigger brand. Zuck and Sheryl had carefully curated this image of Facebook as the Mother Teresa of corporations. I thought this would have presented a darker image. Plus, with how big Facebook aspired to be, I believed that this kind of anticompetitive behavior could have been a red flag for the Department of Justice and Federal Trade Commission.

Inside Zynga, the fight inspired deep heroism and united the company. We embarked on a one-month project, all hands on deck, to build Zynga .com as an independent, freestanding site that could host our own games and have our own social network. Engineers worked around the clock. We had to force people to go home. Moments like this are when your company's true values and culture are forged.

Our board also stepped up. John did what he does best: helped broker bold deals between companies. He walked me into a meeting at Google with Sergey, who had just biked to work and was still in his bike shorts.

"We need Google to lead a $300 million funding round for Zynga," John said. "Otherwise, apps may become captive to Facebook, and that won't be good for Google. Facebook is creating a walled garden that doesn't include Google."

Sergey took over the whiteboard and said something like, "If you're a real platform, you should be able to write 'the founder of that platform is a jerk,' and if that's the most relevant content, that's what should be shown to all users at the top of the feed. The founder can't tip the scales because of some agenda they happen to have for their company at that moment. They're showing their true colors—they're not committed to being a real platform. At Google, we're committed to this open web where content is available."

Google wrote us a $300 million check. This meant we would have a fighting chance to survive independently from Facebook. It's worth noting this came after we had chosen to fight, so it was a massive rocket booster for us all. I was proud that we eventually returned $1 billion to Google for standing by us.

At this point, the teams at Zynga and Facebook had stopped talking to each other. But Zuck and I were still friends and kept texting each other. One night he asked me to come down to Mountain View to try to resolve this, one on one.

We sat in his glassed-in meeting room until 3 or 4 a.m. Zuck was drinking Diet Cokes. I did my best to keep up.

As I recall our conversation, Zuck said, "Zynga is the only company that is capable of being an actual Facebook competitor. You've gotten too big. You have too much of our traffic."

We negotiated a fairly complex deal that maintained our independence while protecting Facebook from the potential for us to move their users to a new platform.

In the clearest sign the two companies had made peace, Zuck made a surprise appearance at our fall 2010 all hands (which was at the Warfield theater in San Francisco), and when I offered to shake hands, he chose a hug.

Two years later, Facebook's 2012 IPO prospectus listed their Zynga dependency as a risk factor. They disclosed: "We currently generate significant revenue from Zynga. . . . If the use of Zynga games on Facebook were to decline, if Zynga launches games on or migrates games to competing platforms, or if we fail to maintain good relations with Zynga, our financial results could be harmed." The filing mentioned Zynga 24 times—more than any other partner or potential competitor.[*]

---

[*] Facebook, Inc., Form S-1 Registration Statement, "Risk Factors," Securities and Exchange Commission, February 1, 2012, https://www.sec.gov/Archives/edgar/data/1326801/000119312512034517/d287954ds1.htm#toc287954_2.

The IPO Is Not Your Friend

Your IPO is usually seen as a win state. But it's actually the biggest challenge to the sustainable future of your company that you'll face. It can destroy your culture and attract people for the wrong reasons. It can take away equity as a currency and motivator and destroy growth mindsets.

After we went public, every value we had started breaking down. "Zynga speed" became a joke.

Being public means you have 100 more jobs, because now every business press pundit is an expert on what's wrong with your company. It made my job a hundred times harder. Michael Dell told me the number one reason he wanted to take Dell private was to own the message again. As a public company, you can no longer be totally transparent with your employees, because you have public disclosure issues. You're constantly being censored by your legal department about what you can say. Your teams are reading stock chat boards and getting narratives from everyone but you.

The only people who benefit from an IPO are those exiting: investors who are cashing out, such as VCs and early team members who are no longer with the company. As a founder, you don't have aligned incentives with your investors. They're not a nonprofit; they're lending you money with a timeline. VC funds commit to a 10-year cycle. That might sound long, but if the fund has been around a few years before investing in you, the time frame could be more like four or five years. Your employees have four-year vesting stock. You've created stakeholders who are not aligned with creating a forever franchise and a forever company.

I learned this the hard way. I had early Zynga engineers who told me they just needed to make $10 million, and then they'd never worry about money again. I arranged a series of private sales for the team before the IPO and got these people their $10 million. They mostly stopped working and left.

Doing a Turnaround When You're Public Is Hard and Not Fun

In May 2012, everything went wrong for Zynga. Our growth engine imploded at the worst possible moment.

For all these years, everything had gone right, and then all of a sudden, the context shifted, and everything went wrong. Facebook's web growth peaked, and the platform changed the newsfeed algorithm and cut off our traffic. We lost 30 percent of our audience in a day. And we were failing in our move to mobile. This was happening six months after we went public. It was a very public face-plant. All of this distracted me and the company for the next couple of years.

We lost sight of our vision, which was delivering on mass-market social gaming, and lost touch with first principles. While it makes sense that we were distracted, it happened, unfortunately, at the very moment when we needed to reinvent social gaming for mobile.

Our stock dropped 60 percent, from $14 in April 2012 to $5 in June, which led to a slew of law firms suing us (as they do for every public company with sudden drops). We had acquired OMGPOP for its hit mobile game, Draw Something, with 15 million DAU; the acquisition looked brilliant at first, but became disastrous a couple months later when the novelty wore off, the turn-based gameplay created friction, and people lost interest and stopped playing. The right thing to do would have been to buy Supercell, which was essentially doing Zynga on mobile. The company was creating brilliant mass-market social games that were gaining traction and taking off. But my board was grumpy about OMGPOP and the lawsuit, so the board challenged me for the first time. Our entire company was distracted by the nonstop negative press about our stock.

We should have come back to our first principles of innovating on social game mechanics and building games we love to play. Supercell did a masterful Proven Better New of FarmVille with Hay Day. Old Zynga would have made a Better Hay Day that was more accessible, social, and fun. Our teams had grown arrogant; they felt that FarmVille 3 had to be a step function better than FarmVille 2, which for them meant it had to be 3D with high-fidelity art. For game developers, 3D is almost always a death trap. It's expensive and slow, which means you can't find the fun quickly. This became a typical failed game project: It burned tens of millions of dollars, and the game never shipped.

We all lost faith, including me. Nothing we were doing was working. And I was burned out. It all felt really hard. It was amazing that I had gone from being named *TechCrunch*'s "founder of the year" to one of *Bloomberg Businessweek*'s "five worst CEOs" in only a year. I was deeply distracted by having to go through all this in the public eye, with the business press watching, and thousands of employees who had once cheered me on because I had made them successful now doubting me. I wondered whether reviving Zynga would require a different kind of leader.

Bing and Doerr came to me with a short-window opportunity to hire Don Mattrick as our CEO, as he was in discussions to be the new CEO of EA. Mattrick had the golden résumé of the game industry. He was a former successful founder who had started his first company at 17 and then went on to become a top leader at EA, building some of the company's most successful franchises. Later, at Microsoft, he had turned around the Xbox. We felt lucky to beat out EA even though he came with an enormous compensation package in excess of $50 million. Despite this cost, our stock went up 20 percent on the news, and the board and employees thanked me.

We hired Mattrick based on a strategy to win the emerging mobile game market with high-production value, heavy-dollar gaming bets. What I later realized was that we needed to do the opposite—to stay close to the metal and get back to reinventing our Zynga engine. Mattrick brought in his own team from the game industry, people with impressive experience who didn't understand social gaming. Unfortunately, they also made us like the rest of the game industry, licensing Tiger Woods and other big IP brands to build the highest-fidelity games possible.

There is a great paragraph in Bezos's 2020 shareholder letter in which he sells Amazon employees on the value of distinctiveness and the danger of becoming normal. He urges Amazon employees to keep alive "the thing or things that make you special."

We saw this play out at Zynga. The new CEO and team's strategy was to launch top-10 mobile hits, which also meant abandoning our mission of connecting the world through games and our playbook of Proven Better New and Golden Mechanics. We were walking away from everything that we had done to be successful.

It was the easy path, but not the right path. Zynga's actual reinvention eventually played out over the next three years and took three CEOs, including me again.

What Will Our Players Thank Us For?

I hoped to stay engaged with Zynga as Chief Product Officer (CPO) and executive chairman and help make Mattrick successful. On a Sunday in July 2013, the day before he started, we met in my office. I planned to share some of our core principles. I wrote a maxim we'd been living by on the wall: "Redlining is less than 25 percent of engineering days committed to future growth." He shut me down: "Mark, I got this. If I need you, I'll call."

No one was asking me to stick around, and now the new CEO was telling me he didn't need me. *Maybe he's right,* I thought. Maybe mobile gaming *was* going to look more like the traditional video game industry. Maybe early Zynga was an aberration and we should let the professionals take over.

The board gave Mattrick free rein to set a new strategy. His plan was to produce eye-wateringly expensive, high-fidelity games on mobile, often with big brands. The board supported his acquisition, for $550 million, of NaturalMotion, a company that seemed to be at the cutting edge of producing high-production mobile games. (Ironically, the board had said no to my acquisition of Supercell for $400 million a year earlier!)

The new management team got Wall Street excited about their slate of new games along with big forecast revenue growth. I had always thought game companies shouldn't give projections to Wall Street about new games, because they're likely to fail. Now all of a sudden, we were committed.

Zynga kept hiring and spending more while neglecting our franchise games, which showed declining App Store ratings and metrics. In the fall of 2014, the Poker team built a whole new version, which they called Poker 2.0. They didn't do Proven Better New; it was just all New. Up to this point, we had the dominant market share in free online poker, but with this new version, our players revolted. Almost overnight, we put World Series of Poker (WSOP), which had been a distant number two, in business. WSOP doubled its user base, while we lost nearly half of ours.

As executive chairman, I demanded that we bring back our original poker game. Zynga did and saw Classic Poker outperform Poker 2.0. While this undermined the management team, it pleased our superfans and propped up the franchise.

Then the team presented a forward business plan for 2015 and 2016 to the board, betting it all on a new slate of games (which they kept delaying, while increasing the projections). I called it a "crack pipe business plan": They were going to drive the company off a cliff, massively increasing spending. But Wall Street liked it; I think they appreciated a grown-up CEO who wore a collared shirt and spoke calmly. Boards and Wall Street don't want to hear truths that are messy. Truth and disruption are more typical of a founder CEO. Confidence and harmony are more typical of a corporate/adult/managerial CEO. (But players and shareholders won't always thank us for harmony.)

By February 2015, the team had further delayed new game release dates and increased spending and projected losses. I pleaded with Mattrick to make big cuts in costs and refocus on core franchises and profitability. Doerr and the rest of the board told me to back off, saying that our CEO was ready to walk. I had turned myself into an expert witness in my own company. I felt closest to the right answer but furthest from operating decision power.

I could have used my voting control to fire the board and replace the CEO, but that felt extreme. At that time, I was thinking we should build a company around consensus. Then, in April 2015, our Chief Financial Officer came to me, worried that the company's plan would fail. I now had data to move the conversation. We had a Sunday meeting with the full board, without Mattrick, and the CFO and I presented where we thought the current plan would land. The presentation was sobering. We were going to lose $120 million that year. If we acted immediately with an alternative plan, we might be able to break even for the year.

This news shocked the board. They asked me to come back to fix the company, and offered $1 in pay. (As an aside, while having the founder/ CEO work for $1 a year sounds great to investors and employees, the reality is that assuming the founder will stay for anything beyond righting the ship without any forward compensation is a trap many boards fall into. That's the reason Tesla shareholders keep voting in favor of huge option packages for Elon. Without any forward compensation, the founder will soon realize they're better off hiring another CEO and pursuing something else with their time.)

We didn't manage the communications with Wall Street well. When I returned, we gave a vague story about our need to retrench around core franchises. We were trying to avoid saying that we had fired our CEO, who had pursued a failing strategy. But that caused even more confusion. The stock fell from $5 to $2 a share. In contrast, in 2019, when Barry Diller fired his CEO at Expedia and stepped in to fix the company, Expedia's stock price went up 7 percent on the news, with further gains afterward.

In my first six weeks, I followed the typical founder turnaround playbook. I convinced Marcus Segal, a friend and early Zynga builder, to come back and help me fix the situation. He took the lead and laid off 18 percent of the company, which was painful but necessary. We closed all of our global data centers. We killed two-thirds of all games and shut down half of our studios around the world.

Longtime Zynga employees were having existential crises, needing one-on-ones, asking us to convince them to stay. I told my assistant that I didn't want to meet with these people anymore; they were going to leave anyway. Colleen, our Chief People Officer, always said, "Focus on the living. Don't waste your time on the dead." Instead, we hired fresh people who loved that we were a $2 stock. They were excited about building a new chapter.

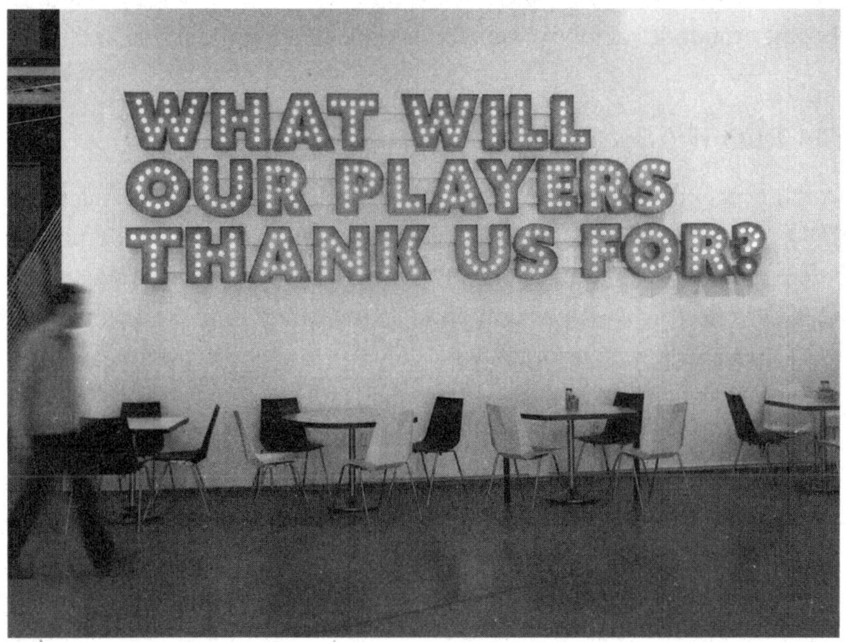

Image courtesy of Zynga, Inc.

That energy shift was key. We went from mourning the company we used to be to being unapologetic about who and what we now were.

I started having weekly product roadmap meetings again and was frustrated to see every team guilty of squeezing the lemon; they were optimizing for monetization and annoying pop-ups, rather than addressing player complaints or creating bold new features. In order to stress a new focus, I shocked my PMs by beginning and ending every meeting with the same question: "What will our players thank us for?" I wanted to rebuild player obsession. So one day, Marcus surprised me by having the question posted as a giant neon sign in the lobby. We went back to first principles and focused on quality. We read our App Store reviews and tried to build what players liked instead of what would drive revenue. As we did so, we started to see improvement in all metrics across our core franchises.

I focused on Words With Friends because it was one of our most valuable franchises. It was projected to go from $120 million in 2014 revenue to $79 million for 2015. By 2016, we had grown it to $180 million. Even more important, the game generated more profit ($100 million) than the entire previous revenue forecast! This proved one of the core theses of the Zynga turnaround: that the economics of forever franchises were massive and powerful.

The Truth Will Set You Free

In fall 2015, our CFO came to me and said that he thought he should be COO. I told him he was a good CFO but needed to dominate the position before considering anything bigger. He told me he had COO opportunities elsewhere, so this was the right point in his career to make a move. This was a stressful moment for me with only three weeks left until earnings.

"This isn't going to look good," he said.

"Yeah, you're right, but that's just where we are," I replied.

There's a certain authenticity that comes when you're completely beaten down. There's nowhere further to go. That quarter, I shared with investors: "Our CFO resigned to pursue better opportunities." No one says that! They always say "for personal reasons" or some corporate lie. It was very freeing to tell the truth. I was still nervous about the earnings call, but nobody asked anything. At the end of the call, Michael Pachter from Wedbush

Securities said, "Mark, I see that your CFO quit. Can you tell us what you are looking for in a new CFO? Maybe we can help you source more people."

When you are transparent, and your authenticity reads clearly, no one questions it, and they move on.

Hiring the Right CEO

While the turnaround was clearly working, I wasn't having fun. I was playing defense, trying to cut costs and marginally improve games. The board had become this legal governing body with committees—corporate and cumbersome. And of course, nobody was paying me anyway. Later in 2015, Bing introduced me to Frank Gibeau, who had just left EA. Frank turned out to be a kindred spirit who loved games and liked being in the details. Equally important, Frank passed my beer and sushi test. If you're going on a big journey with a CEO or board member, make sure you enjoy hanging out. We convinced him to join the board. He started showing up a day or two a week, joining my C-staff meetings. His biggest win as a board member was fighting for us to save CSR2, which was a sequel NaturalMotion created to CSR, its popular racing game. We were cutting the company to the bone, and everything New had to go. He made a compelling case that this wasn't really New and would be an attractive upgrade to what he viewed as a forever franchise. We went with his recommendation, and he was right, revealing some of the signs of a winning CEO.

I asked Frank to be CEO, and in March 2016 he stepped in and I moved to the executive chairman role, this time in a full-time capacity. Frank was committed to reconstituting what we called CPM, central product management, our core strength. He believed in the original Zynga DNA: franchises, roadmaps, engineering days, Bold Beats, and "all New fails." He knew how to institutionalize these DNA strands with practices such as making Bold Beats part of the quarterly OKRs for every studio. He believed my challenge—"What will our players thank us for?"—was vital to the turnaround.

Frank coined the term "forever franchises" and made these a core part of the company strategy. He took this narrative to Wall Street, which started using the same success metrics for us—a turning point for the company.

He brought in a great team and pursued mergers and acquisitions to buy more forever franchises. He saw our competency as being great at LiveOps (running games as a service) and regrew the company footprint through that narrative.

In May 2018, with Zynga in a strong position and the team focused on our players, I swapped the exec chairman role for a non-executive chairman role and went back to building and incubating products.

In January 2022, Take-Two Interactive came to us with a merger offer. The company needed Zynga to fill a hole created by the continuing delays in launching *Grand Theft Auto VI*. (As I write this, in March 2026, it's still not out!) The merger was a successful one; Take-Two's stock was at $169 when it bought Zynga (for $9.60/share, nearly five times what the price was when I came back as CEO). In October 2025, Take-Two's stock was at $258, largely because of how well Zynga has done. Frank is still leading Zynga as part of the Take-Two team.

While I think of Zynga's outcome as a B+ because of what the company and social gaming *could have* become, I'm happy there was a successful exit for shareholders and employees, and that the company and its franchises are still going strong. To this day, I still believe that all of us are looking for permission to play. It's amazing to me that the games industry generates $280 billion and has three billion DAU and yet remains mostly empty calories. If Zynga had managed to stay on our mission of connecting the world through games, perhaps today games could be a good use of people's time.

Founder Mode over Harmony

Zynga's stock went up when we hired Mattrick, down when I came back, and up again when Frank joined. There was a view on Wall Street that founders aren't their friends. (Maybe that's starting to turn; some people note that founder-led companies have outperformed the market.) But I was the best friend Wall Street had in the boardroom all along. When I came back, I did what the grown-up CEO was supposed to do: massively cut costs. I bought back $200 million worth of stock at an average of $2.60 a share. The analyst for our biggest public holder yelled at me for not buying it back at a lower price, because it had sunk further to $1.75. Being a founder, like being a parent, means that you take the hits for the hard decisions but don't usually get thanks later for being right.

When I came back to Zynga, I was a bit of a lapdog, trying to please my board and Wall Street. Zynga was an amazing asset base, even as a $2 stock. We had a billion in cash and franchises that needed to be fixed but that, once they were rightsized, had the potential to generate $300–$400 million per year. I look at Michael Saylor and admire how he stuck with MicroStrategy, an old enterprise software company with no growth prospects, and acted entrepreneurially by taking his free cash and transforming his company into one of the biggest public market bets on Bitcoin. I had a platform with great people and scale, and I regret that I didn't stop time and view it as the opportunity it was, instead of seeing it as a burden that needed fixing. The key lesson is to value our instincts over consensus and harmony. Be bold and go for it all the time.

Looking back, I wish I could have acted with more conviction. When things went bad in 2012, I wish I'd said, "Maybe I need a break, but I'm the right person to fix this." I wish I'd had the courage to stand up to and potentially fire my board when they wouldn't let me buy Supercell. And I learned that when you do finally agree to come back and fix things, negotiate for all you deserve *before* taking on the job. I should have replaced many of my board members with people who could stomach upheaval and change.

So much of this is about your perspective in the moment. Mark 2012 and Mark 2015 thought he was dealt a bad hand. Mark 2025 thinks it was a great hand; I just wasn't ready to play it.

TL;DR

Your number one job as founder and CEO is to be right. Scale isn't the objective; quality is. Both being right and quality come from staying close to the metal—asking, "What will our players thank us for?," and staying focused on the answer, rather than getting distracted. When you do have to scale, develop processes and practices that get people to do the right thing when you're not in the room. Focus on growing and hiring leaders and staying true to your instincts about product and people.

8

Landing on New Planets

Imagine what it would actually be like to live on Mars.

It would be *hard*.

Imagine building products on platforms that we don't use in our daily lives yet.

Also *hard*.

Building for new platforms—as we did for the web and mobile—is like landing on a new planet. Each new planet develops its own languages, customs, and native inhabitants. The pioneers who first inhabited these spaces were crawling across broken glass. The friction in using these spaces was enormous, the benefits weren't obvious, and the experience was worse than what these were supposedly replacing. But these pioneers saw something the rest of the world didn't yet realize.

We are living in a crazy time. And I feel the deepest FOMO of my career. Ninety-nine percent of founders are feeling the same way. None of us founded OpenAI. But one of us may unlock its greatest value for the world.

Today, we have so many reasons to be optimistic. We're on the verge of five new planets: AI, robotics, space, crypto, and the metaverse. AI is massively reducing the time and cost to turn ideas into apps we can test. We have line of sight to humanoid robotics. It hurts my brain to imagine what life will be like when there are nine billion R2-D2s following us around. I often turn to the TV show *Black Mirror* to help me imagine the future, even though my own belief is that these innovations will enhance our lives rather than drop us into a dystopia. Space exploration is the most foreign emergent industry to me. I can clearly see the business potential in Starlink satellites spreading wireless broadband around the planet. However, I struggle to grasp the other opportunities, like space pharmaceuticals, mining the moon and other planets, and solar harvesting. And we've almost taken crypto/blockchain for granted, even though it offers the promise to reinvent money and contracts after 2,000 years. The metaverse has gotten a bad name, because it turned out none of us wanted to live inside immersive goggles, but I'm still a believer in the underlying concept. The lines between virtual and real are blurring in ways that will reinvent industries. All listings businesses, from dating to jobs to all e-commerce, will be taken over by agents as they get better at talking to one another. This will revolutionize (and possibly, obliterate and replace) today's existing database-centric businesses, such as LinkedIn and even Meta.

In this chapter, I try to address the elephant in the room: How do we apply the philosophies and frameworks in this book to go after emergent new markets—these new planets? How do we apply Proven Better New when there is no Proven? How do we engage with these new planets, in both small ways and big ways? How do we test without fully committing? What lessons from past shifts will repeat themselves? How do we start living on these planets early, while it's still hard and doesn't make economic sense? And how do we discover our own passions amid this chaos and give ourselves permission to be curious?

The toughest question for us all: With technology moving faster than anyone can track, with the world being reinvented daily, what do we do this week?

Part I: New Planets Break Our Frameworks

The core problem is that it's so hard to approach these new planets when nothing is Proven. You could just throw Proven Better New out the window and say, "That framework works only in a mature market where Proven has been established." But that would be a mistake.

You could also easily misapply Proven Better New. That's what I did with SMS Taxi. As I mentioned in chapter 2, "Instincts Versus Ideas," in 2003, I registered smstaxi.com because I had the instinct that we should be able to order a taxi through our phone. That was obviously right, beyond what anyone imagined. However, I wrongly applied Proven to generate a losing idea: ordering a taxi. At the time, I was trying to apply Web 1.0 principles of simple ecommerce to what ultimately became a Web 2.0 opportunity. I was misapplying Proven Better New when I said what was Proven was taxis.

The reality of Uber was that Proven had to be applied more broadly to the concept of a taxi, as a paid ride from point A to point B. Better was ordering through your phone and seeing the status on a mobile Google Map (which didn't yet exist). The key component was the New: Anyone could now be that taxi, which both lowered the price and massively expanded the availability of a ride from point A to point B. And I/we couldn't have known that idea would ever work. If we had tested the Uber concept before people lived on mobile devices, focus groups would have overwhelmingly rejected it. If asked, "Would you get in a stranger's car?," most if not all people would have said no. The concept wouldn't become obvious until we lived this mobile-centric lifestyle, in which we value convenience, availability, and low price so much that we overcome trust issues with strangers.

Web 1.0, 2.0, 3.0 Defined

Web 1.0 (1990s–early 2000s): Read-only internet. Static websites you could browse but not interact with—like a digital brochure. Think early Yahoo, news sites, and company home pages.

Web 2.0 (mid-2000s–now): Read-write internet. Users create content, not just consume it. Think social media, user reviews, YouTube, and Wikipedia.

Web 3.0 (emerging): Read-write-own internet. Users own their data, identity, and digital assets through blockchain and crypto. Decentralized, no middleman. Think NFTs, crypto wallets, and Decentralized Autonomous Organizations (DAOs).

Lessons from Previous Planets: Web 1.0, 2.0, and 3.0

In every platform shift, there's a first wave of services and companies that predictably emerge. We see the "big iron" first; originally, this was the mainframe. In Web 1.0, which included Amazon, eBay, and Google, everyone predicted that ecommerce would replace catalogs and stores. This prediction was so obvious that companies got overinvested before it even made any sense. Pets.com is a prime example.

There's a second wave, though, which you can't imagine until you've lived on that planet. We didn't know that eBay and Craigslist would be runaway hits until we were native to Web 1.0. These unpredictable breakouts demonstrated the Jevons paradox: When a resource becomes nearly free, demand can explode exponentially and in unexpected ways. When listing items online became free, people listed everything. I remember being shocked that I could sell my useless Sony Aibo robotic dog for a profit. Craigslist expanded beyond traditional classifieds to "take this couch for free" offers and roommate matching. My partner, Hilary, found her roommates, who eventually became some of her best friends, this way.

The first wave of Web 2.0 brought Myspace, Facebook, and LinkedIn, but we never imagined Uber and Airbnb, which emerged in the second wave. What we couldn't have predicted about Web 2.0 was the role that trust would play. As we moved to a web of people, not pages, and users had online identities that connected to their real selves, trust became the critical factor. Facebook and LinkedIn succeeded where my social network Tribe.net failed because they got trust right. As noted, I didn't anticipate how deeply people would care about privacy and control over their identities when moving online. But once they became comfortable with that, ironically (and shockingly) they were willing to trust getting in strangers' cars and staying in strangers' apartments. That would have seemed absurd during Web 1.0.

Web 3.0: A Case Study in False Signals

Web 3.0 is a textbook example of a technology innovation that launched while still searching for true signal with consumers. Experienced investors poured billions into blockchain start-ups while ignoring the lack of genuine consumer traction beyond day trading. VCs thought games would be the first Web 3.0 use case. The thesis was logical: Players spend money on virtual items that get trapped in games; blockchain would let them sell these assets. The problem was that nobody was going to play a game because it was on the blockchain. Blockchain might increase monetization, but that doesn't make the game more fun.

Nothing else worked either. I attended a VC lunch where a blockchain payroll founder's primary evidence of adoption was that his own company used his system. These are "drinking the Kool-Aid" moments, when we're so excited about the future that we misread signals.

Compare this to true signal: Napster had millions of kids sharing music in months. Facebook signed up 20 percent of each school in the first week, 80 percent the following week. In both cases, founders needed cash to buy more servers; that's always a buy sign. True signal always shows in user behavior, not just technology capability.

Today, Web 3.0 appears to be making the inevitable, eventual comeback, but in a form most people couldn't have predicted. As of February 2026, Bitcoin has a market cap of $1.3 trillion, but this is all because it is now considered a legitimate form of digital gold, or value store. It has yet to show any real traction around its original intended use, which was as a blockchain and payment system. In fact, anyone who ever bought anything with Bitcoin is probably sorry today!

Part II: Adapting and Redefining Proven for New Platforms

When we move to a new platform, nothing is Proven—as I've defined it—for this audience on this platform. We need to define what "Proven" means much more loosely in this context. Let's call it "Fuzzy Proven." A lot of what we're doing is applying mechanics that have been proven somewhere else to generate a good shot on goal, meaning a product people are likely to want. Hopefully, we do this in a way that feels native to the new platform.

As we saw with my SMS Taxi miss, you have to get to the atomic unit of Proven. Deconstruct the experience. What is Proven about behaviors? What are the core mechanics that transfer?

For new platforms, the customs, mores, and practices aren't yet established. When we first landed on Planet Mobile, the kind of interface people wanted wasn't invented yet. If you just applied what was Proven on web, you wouldn't have been considering what it meant to be native to mobile. The need to economize everything, down to the pixel, was the beauty and challenge of mobile.

And yet, while we needed to reimagine the experience, we also needed to hang on to the right Proven. So many people approached mobile as a license to reinvent every aspect of their product. Most failed because they ignored the nuance of Proven mechanics. You need to determine what is Proven and why.

With Hay Day, Supercell delivered a masterful Proven of FarmVille. Supercell didn't reinvent; the company reinterpreted the game for mobile. They didn't change what already worked (recall the half-circle driveway and the art of FarmVille described in chapter 4, "Proven Better New"). Supercell got a PhD in farming simulation games and picked the right Proven elements to adapt for mobile—making crop times shorter, simplifying the menus, and magnifying the delight of farming through touch screens rather than mouse clicking.

People who master this Fuzzy Proven, as Supercell did—who go to the right atomic unit of what is Proven—have a huge advantage on a new platform. But you must get Proven right. You must understand what is state of the art and why, to apply to your category. That's why I love the quote from the French writer Jean-Baptiste Alphonse Karr: "The more things change, the more they stay the same."

In April 2023, I tried (and failed) to do Proven Better New to generate early traction and learning on Planet AI. ChatGPT was just taking off but didn't have a mobile app yet. My team and I thought there was an opportunity to move quickly and launch a better AI chat interface centered around personalized agents. We saw Poe, created by Quora CEO Adam D'Angelo, taking off and thought we could do Proven Better New by giving people a trusted friend. The effort failed before we even launched because we underestimated what was Proven in Poe. We thought Poe was just a better mobile UX. But when we compared our chat query responses to Poe's, Poe's were better. We couldn't get to Poe's Proven, much less Better. The main benefit of this exercise was that it got our team native to AI much faster.

In another, more recent example, in September 2025, my Erth.AI team released Garden Party, a web game on the popular chat platform Discord, to show off our moddable game engine, Stem Studio. The concept was to put out a game that players could remix into their own variations—think of it as vibe game creation. We chose to build our own version of Grow a Garden, a Roblox game that was a massive hit, achieving over 100 million DAU and, by my estimate, $6 million in gross daily revenues. That's heat!

Our Garden Party game is a case of Fuzzy Proven, because the original game blew up inside Roblox, which didn't mean it would generate heat on web or mobile. However, it offered a canvas for us to show off the capabilities of our modding engine with new features such as "Create with AI." The process was far faster than inventing a new game, which would introduce risk and delay testing for signal on what we cared about. While the game wasn't a hit, it was useful for our strategy to test our idea for easy modding with real players, searching for experiences they love. My hypothesis is that people want to experience the magic of "vibe creating" rather than just consume content.

Whether or not any of these early attempts succeed isn't the point. They are all part of a necessary process to get native to AI.

Getting Native

In 1982, my dad tried and failed to get native to the coming PC revolution. He bought an odd early PC called a Kaypro II, which was an ugly-looking machine with a terrible form factor. He spent months re-creating all his writings and files on the system. But the machine kept crashing, and eventually it became a $2K, 26-pound paperweight. The experience burned him so badly that he avoided computers for the rest of his life. He even had his secretary print out his emails, and he wrote replies on the printouts with a red felt-tip pen.

I trace my internet native origin to 1993, when I worked at TCI (the biggest cable company). You might recall that, in chapter 1, I mentioned how my tech journey began with reading *Microcosm* by George Gilder. Gilder also played a role in my landing my dream job at TCI, which I saw as the future on-ramp to the information superhighway (which became the internet). I mentioned Gilder in my 15-minute interview with CEO and legend John Malone and learned he was meeting him that same night for dinner.

TCI had recently announced its digital set-top box, which was supposed to deliver 500 channels of TV and every interactive service anyone had ever dreamed of. When the TCI president asked me why they needed a second MBA (in a company of 20,000!), I said that there must be tons of opportunities coming in and I could analyze them. When I was hired, I learned the reality: Every deal other than buying cable systems was tossed into a cardboard box called "non-cable," and my job was to fish through it and work with TCI's Liberty Media division to bring these ideas to life.

This job was like a hunting license for going native. One of Liberty's first experiments was proving people wanted video on demand. This was actually a terrific example of quickly offering a Minimum Idea State, rather than building a version that was expensive and right. We started with a "roller skate test." When viewers requested a video, someone at the cable company literally roller-skated over to racks of VHS machines, inserted a tape, and pressed play, while at home, viewers felt they were experiencing a magical future.

One early insight from this experiment was that every new digital platform would first get traction with porn. We also had an in-hotel room on-demand movie service called SpectraVision that made all of its money from porn. No surprise today. The porn insight is now universally known. I do remember going to the Consumer Electronics Show (CES) in the 1990s because the porn trade show, held at the same time and place as CES, featured the most cutting-edge new web technologies.

So much of going native is about playing with the new technology long before it's "worth the time"—which means you follow pure curiosity about what is possible without stopping to ask whether it will be easy to do or even work at all. We all become hobbyists. It's kind of beautiful since none of this is a real business, so we have the chance to find the "edge" of where these applications are before real companies and serious engineers ever bother.

For me, this has happened many times. I remember in 2006, before launching Zynga, I spent hours on setup to play *Rise of Nations* with my niece and nephews online. It took four PCs for two of us to play. Two PCs had the game via (I think) GameSpy Network, and two machines had a live video connection, maybe over Webex. It was really fun, and we did it exactly once.

By enduring those moments of resistance, however, we build muscle memory around each friction point. We start to know the problems so intimately that we will be the first to spot them and maybe go bananas when we see them finally being solved.

I became native to social networking by building Tribe, which helped me recognize Facebook's potential. By the time Facebook's Matt Cohler told me that the company was opening up the platform, I knew we could connect people socially through games. And I became more native to Facebook by building apps on the platform.

Sometimes we can completely miss opportunities to get native to a new platform and then suffer having to catch up later. I remember sitting in the car while Reid and Sam Altman were debating how to make AGI (artificial general intelligence) safe for humanity. I couldn't see how we would ever get to AGI, and I definitely didn't worry about saving the planet from it. Eventually Reid showed me GPT-3 and how it could make up lightbulb jokes about me that were good enough to sting. Then I got it.

Right now, as we are at the beginning of this AI revolution, there are so many ways we can get native. For example, I loved the internal memo Shopify CEO Tobi Lütke shared with his team. He said, "Before asking for more headcount and resources, teams must demonstrate why they cannot get what they want done using AI." That is the way to get everyone on your team to understand you've landed on Planet AI.

One big way I've been getting native to AI has been to use Waymo's self-driving cars instead of Uber—not because it's always more convenient, but because it's far better having no stranger in the car. We get our privacy back. I suggested to Sergey Brin, cofounder of Google and Waymo owner, that the company is undercharging by pricing against UberX and not Uber Black. And sometimes, I still feel that the AI is like a child that needs more training—for example, when it gets confused and stuck behind someone waiting to park or construction vehicles. Maybe Waymos are too polite?

We All Need to Be Hackers

The hobbyists who get to solutions early generate massive value. They do so by hand, in brute-force ways, often with creative workarounds, but they get there first. That was the case with my first product, FreeLoader, which crashed all the time but, when it did work, delivered a magical experience long before anyone else could get there.

I remember when we demoed Zynga Poker to Steve Jobs, and he thought it was fake—that is, not a live game. He yelled at Scott Forstall, Apple's head of software, that he didn't want to see demoware, only working live code. But the game was real. Our brilliant cowboy engineer, Scott Dale, had hacked Facebook and Myspace web APIs to work with our native iOS mobile app. Once Jobs realized this was real—in fact, I challenged him to chat with the players if he wanted—he flipped: "My son will play that and love it."

These hacky little innovations become obvious later. I know this is inside baseball, but Nat Friedman, former GitHub CEO, built an early GPT-3 integration with GitHub. It was mostly useless (by his own account), but it made every engineer dream of a coding copilot so much that they were still in awe of the vision. That's the hacky mentality you need when nothing is built yet to live on Mars, or mobile, or AI. Now AI code editing is a massive industry. Often, being first with these hacky solutions can lead to owning a future market.

At the same time, my engineers have always been skeptical of small early innovations because they're so unimpressive and easy to build. The engineers always say, "We can build that." But they don't, and they won't. At Zynga, we had the opportunity at the outset of the App Store to buy Test-Flight for $500K. We didn't buy it or build it. Eighteen years later, after Apple acquired it, TestFlight remains the gold standard for mobile app testing. Is it important to the app ecosystem today and worth owning? Absolutely.

Permission to Be Too Early

What do you do when you see technology innovation without any clear consumer signal? I call this zone "permission to be too early"—a state where it's acceptable, even advantageous, to "waste time" on small, narrow, yet interesting products.

This zone offers some key advantages: less competition; easier experimentation; more forgiving consumers; and hobbyist, insider communities eager for new contributions. Today's prosumers—power users who both consume and create—form around AI, with splinter groups emerging in vibe coding, agents, and AI gaming.

The likelihood of getting timing perfectly right is very low. You'll almost certainly be too early or too late—so being too early is preferable. Assume your initial attempts will miss the mark, then iterate rapidly.

People say Zynga missed mobile, but actually, we were too early. We built an 800-person division and acquired Draw Something, the most popular mobile game, while our competitors stayed focused on Facebook. When we made mobile games, we were launching titles that at best would do $50K a day in place of web games that were generating $500K a day. We were trading dollars for pennies; it was the classic innovator's dilemma.

When mobile emerged, Bing Gordon said we should be betting 10 percent of our market cap every year on the next big market, which was clearly mobile. In fact, we see Meta doing something similar today to catch up in AI, even though at times this approach upsets Wall Street. As I mentioned in the last chapter, our big miss was in not buying Supercell, which ultimately became the leader in mobile games. We had a handshake deal with the founder, Ilkka Paananen, for $400 million, which probably wasn't even 10 percent of our capitalization, but my board said no. If we had given ourselves permission to truly be too early, we would have made different bets. We would have continued to invest in mobile, and we would have bought Supercell.

Finding Your Curious Edges

One of the most practical ways to start living on new planets is to find your curious edges—places where your personal life intersects with emerging technologies. Your curious edges often come disguised as life problems. They present as challenges, but they're actually opportunities to get close to planets before they're inhabited.

My example is gene editing for my son, Wyatt, who has a gene deletion. We've been told that a medical solution for kids like him is 50 years away. What if we could accelerate that to 10 years? This personal challenge has led me to want to go deeper into understanding genetic engineering for the millions of people born with rare gene deletions.

Eric Lefkofsky, the cofounder of Groupon, had a similar experience. When his wife got breast cancer, he found what he wanted to solve next. Why couldn't breast cancer be detected earlier through genome sequencing? He got to the edge of early detection, pattern matching, and genomic analysis. That challenge became Tempus, a precision medicine company that brings data and AI to health care. He took the company public in 2024 and now, in 2026, it serves millions of people and has a market capitalization of $8.6 billion.

The point is to get native in one corner of a new planet where you have genuine motivation to keep exploring, even when it's hard and doesn't yet make economic sense.

Increase Your Surface Area

One of the most practical ways to land on new planets is to expose yourself to people who are already there building things.

One night in August 2025, my friend Martin stopped by with his friend Josh, a magazine writer, former *Wired* editor, and award-winning filmmaker. Josh showed us an application he'd built without writing a line of code—a voice bot that calls people and captures their stories. When he demonstrated it by having it call me, the experience felt magical. The "Josh.AI" asked about my best story, and when I paused, thinking, it pivoted naturally: "Have you ever been attacked by a wild animal?" I actually had, so I started telling that story. My daughter Georgia was mesmerized.

I realized afterward that this random guy who walked into my house is living life at the speed of play. As a writer, he is supposed to be the next roadkill. Instead, he's playing with AI, going deeper with his craft, and having fun. Josh knows what makes a great story, how to find it, and how to share it. His instincts come from his taste, which he's calibrated from his many years as a storyteller. He'd built this product in days by asking AI the right questions and pasting together APIs.

What struck me wasn't just the product itself. It was that this random encounter opened up an instinct vein for me around voice. I realized voice might be the killer interface for AI apps in ways I hadn't fully considered. Will we all move to talking instead of clicking and typing? Josh made me wonder whether the change will be more profound than I'd been thinking. (In fact, around this same time, Reid put out a thought piece saying that he had been "voicepilled.")

That's what happens when you focus beyond your own projects and see what other people are building. You don't need to adopt their exact idea or solution. But seeing someone else's approach to a new planet can unlock something in your own thinking—a new angle, a missing piece, or even an instinct you didn't know you had.

After this, I realized I needed to find more Joshes—to get out of my office more and expose myself to people who are building interesting things, not theorizing about them. Focusing outward can be uncomfortable for many of us but is necessary to land on new planets—not to wait for inspiration to strike in the vacuum of our own offices, but to actively seek out the people who are already exploring.

Part III: Three Approaches to Landing on New Planets

There is no single playbook for landing on new planets. However, it's useful to learn from the vastly different approaches taken by the most prolific founders and investors. That's why I interviewed three of the pioneers I admire most. Reid Hoffman has nailed every phase of the internet and most recently was one of the key enablers of OpenAI. Cyan Banister's pursuit of weird has led to gold, and Neal Stephenson invented the term "metaverse" and is still making it come true 30 years later. (When we asked ChatGPT how it would characterize my approach to making products, the chatbot called it the teaching hospital and said, "Get your PhD in the mechanics, then adapt.")

Reid Hoffman: People, People, People

I first met Reid in 2000 when he was at PayPal, and we connected around politics. We've been thought partners for decades, collaborating around the beginning of social networking, and I've gotten to see how he approaches new planets by both working with him and watching him operate.

Reid's whole approach to landing on new planets is network-centric.

"Engineers think 'people, places, things,' but it's actually 'people, people, people'"—places and things show up way down the list, Reid says. "What people most care about is people."

This people-first lens helped Reid see that social networking would be massive before almost everyone else did. As founders do, he created LinkedIn in the image of himself (master networker). LinkedIn wasn't about replacing Monster.com, the leading online job listings business (even though it did, by a long shot); it was about people wanting to connect with people in a professional context.

Reid leverages relationships that he's been investing in for decades as "tentacles" that bring back signals about emergent new technologies and markets. Once he identifies new opportunities, he uses his network to rapidly prosecute and hyper-scale new market-leading companies. OpenAI is the biggest and latest example. Reid played a fundamental role, both writing the first $10 million check from his foundation before OpenAI had any commercial potential and then brokering multiple massive capital deals with Microsoft that enabled OpenAI to become what it is today. He was a pivotal bridge in his dual roles as a board member of Microsoft and OpenAI.

Reid's mission around AI is to amplify humanity. And he has the best virtual Rolodex to pursue this, immediately connecting him to the most relevant native experts. He teamed with Mustafa Suleyman, one of the founders of DeepMind, to start Inflection AI, a pioneering LLM company that later entered into a multibillion-dollar commercial agreement with Microsoft. More recently, Reid has partnered with one of the world's top oncology researchers, Dr. Siddhartha Mukherjee, to create Manas AI, to cure cancer and rare diseases. (I'm happy to be an early investor.)

The lesson here is that we should all build our own networks and expand the reach of our own tentacles. Reid invested in relationships and built infrastructure years before the payoff was clear.

Was XiaoIce the Proven for ChatGPT?

In 2021, long before OpenAI launched a GPT chatbot product, I sat in on an amazing lunch discussion between Bill Gates and Reid. During a break in Reid's Masters of Scale conference, Bill talked about XiaoIce, a Chinese chatbot app launched in 2014, based on early Microsoft AI, which, at the time, had 150 million users. XiaoIce had taken off because people loved having someone they could talk to any time about anything. I trace back to that as the first actual Proven behind ChatGPT. XiaoIce, launched by a team of Microsoft Bing researchers, became a sensation in China for its human-like conversation abilities; people were talking to it as if it were a friend. I think Reid and Sam must have connected the dots and realized this was the killer consumer use for their AI.

Reid Is an AI Power User

Reid is constantly experimenting. He has an interactive video avatar trained on his mannerisms that can have conversations as if it's Reid himself. It has its own speaking tour. He co-wrote a bestselling book with GPT-4 called *Impromptu* about how AI can amplify humanity across education, business, and creativity. Another project of his is taking classic works and rewriting them for modern contexts, like Machiavelli's *The Prince* for Silicon Valley. It's not just about the output; it's about the process of working with AI to understand its capabilities and limitations. "Fucking use it to see the future," he says.

Reid told me he learned this lesson with Tesla. He thought he understood electric vehicles intellectually: "It's moving cars from mechanical engineering to a software engineering paradigm." But when he actually drove one, he said, "Five minutes in, I started having new ideas about those paradigms."

The Seven Deadly Sins Investment Filter

Since it's all about people for Reid, it makes sense that he makes investments based on fundamental human needs—or what might be thought of as the "seven deadly sins investment filter." As Reid states it, "Facebook is vanity/pride, Twitter is wrath, LinkedIn is greed, Zynga is sloth." He didn't list the rest of the seven sins. But his point is there has to be an innate, selfish behavior that makes people use the service. This gives him a framework for predicting what will scale to hundreds of millions of users. If it doesn't tap into a basic human urge, it won't achieve massive adoption.

Bits Versus Atoms

When choosing which new planets to explore, Reid has a clear hierarchy: "I generally start from a state of bits versus atoms. Investing in bits is much, much better, even on impact."

This explains why Reid went all in on AI (bits) rather than robotics or space exploration (atoms). Software scales infinitely; hardware doesn't. You can iterate on code in hours; manufacturing takes months. As he puts it, "I'm glad that people invest in atoms, and we should create systems so more people invest in atoms." But for most entrepreneurs, especially those just starting to explore new planets, bits-first gives you faster iteration cycles and cheaper experiments.

Reid, however, distinguishes between "economic investments" (following his rules) and "impact investments" (breaking his rules for things he wants to see happen). For instance, he backed some nuclear fusion companies not because they fit his bits-first philosophy, but because he wanted those technologies to exist in the world.

I walked away super energized from my deep dive with Reid. I had never heard him refer to his approach as "people, people, people." That made so much sense. It was so obvious, but somehow, over 25 years, I never saw things that clearly and simply. He finds such interesting ways to get native. He pursues any hobby or passion by applying AI. And he has the discipline to ask, not just whether something is a cool new product, but which "deadly sin" or core human need it addresses.

Cyan Banister: Pursuing the Weird

Cyan has consistently been early and right. She founded Zivity a decade before OnlyFans and has been a founding investor in major contrarian winners such as SpaceX, Anduril, Uber, and DeepMind, in addition to other winners including Niantic, Postmates, Carta, Thumbtack, Flexport, and Affirm.

I've known and admired her for decades, especially for her courage and willingness to promote contrarian views (most of which I have agreed with). And I love her pursuit of the weird and how deep she'll go to pursue her own curiosities.

Cyan's own start-up experience shaped her investment philosophy. In 2007, she built Zivity, an early version of OnlyFans that allowed creators to post nude pictures for money. Like Tribe, this was a classic case study in instincts versus ideas. Looking back, she sees lots of reasons it didn't work, starting with timing: "We were the General Magic of this sector. We had all the right ideas but the wrong timing. Timing really matters, and at the time we launched, Facebook had fewer than 100,000 users, and Myspace was a thing."

Cyan is used to being not just a pioneer but a pariah. "If I showed up at tech events there was this contingent of people who worked very hard to try and make my life a living hellscape for daring to be a start-up in this space." She was really crawling across broken glass trying to do this so early. As she says, this wasn't a company you could call home and brag about. She couldn't get funding or employees.

"We invented crowdfunding," she says. "We didn't realize it at the time." They called it "voting" (Kickstarter later called it "backing"). As I learned at Tribe, being right about the future isn't enough if you can't execute in the present. But being early gives you pattern recognition when similar opportunities emerge. That's why Cyan is comfortable funding weird ideas that seem impossibly early. She believes that the person who sticks with a vision longest—who keeps iterating on the execution while everyone else gives up—often wins when the timing finally aligns.

Today Cyan is a general partner at her own firm, Long Journey Ventures. She focuses on pre-seed and seed rounds, often being the first institutional investor. She builds a "secret network" of engineers and artists who leak weird ideas to her. When everyone else is saying, "This is insane" or "The timing is wrong," she moves in. Her $100K "Wayfinder" checks let her move at angel speed with high conviction. If she sees something strange that others dismiss, she can fund it immediately without committee approval.

Cyan looks for what she calls "divergent minds"—people who ask questions she's never been asked before or paint pictures of a world she's never seen. She's drawn to founders who make her uncomfortable, who propose ideas that sound absurd. Her brand is being the person you call when you have a weird idea. She wants to be known as the "Ghostbusters phone number"—when someone sees something strange and doesn't know whom else to call, they call Cyan.

Cyan's approach works because she's not trying to have a high hit rate; she's trying to find the one idea that reshapes an industry. She caught Uber by seeing the regulatory constraints (taxi medallions) as artificial bottlenecks that could be removed.

What makes Cyan unusual is that she treats curiosity as a systematic practice. She literally schedules time to be unproductive. She'll turn off her phone for eight hours and follow weird breadcrumbs of coincidences; if she sees a strange sign on a wall, she'll go there. She uses dice to decide where to spend her next hour. She says she treats everything like a game.

This approach led to one of her best investments when she was at Founders Fund. Sitting in the lobby of the Four Seasons, she noticed wi-fi names including "Garrett Langley's iPhone." She googled him and discovered he was at Y Combinator building Flock Safety (license plate recognition for neighborhoods). She walked over and introduced herself, and Founders Fund nabbed the last allocation in the round (and co-led their next round!). The company is now worth over a billion dollars. Most investors think you find good deals through meetings and networking. Cyan finds them by wandering and wondering.

How Cyan Gets Native

In many ways, Cyan approaches new planets like a method actor. She doesn't just research industries; she lives in them. During the pandemic, she tried to manufacture a simple enamel pin in America—just out of curiosity. After weeks of delays and complications, she gave up and tried Alibaba. Within 48 hours, she had a prototype from someone who spoke perfect English. Within a week, she had finished pins.

The experience wasn't about making pins. It was about understanding the manufacturing gap that needs filling. It led her to invest in companies trying to build an "Alibaba of the Americas"—automated manufacturing systems that could compete with Chinese production. She gets native by fully immersing herself, not just analyzing from afar.

Neal Stephenson: Build the Future You Imagine

It's amazing to see how many predictions Neal made decades ago in his seminal novel, *Snow Crash,* coining the term "metaverse" and popularizing the word "avatar." He's written more than a dozen novels, won the Hugo Award for best science fiction novel, and was the Chief Futurist at the augmented reality start-up Magic Leap. His ideas have had massive influence in the real world. Sergey Brin calls the book one of his favorite novels and the developers of Google Earth cited it as an influence.

What makes Neal unique as a planet pioneer, though, isn't just his ability to see the future; it's his willingness to build the infrastructure for his future visions long before they're even real.

I've gotten to know Neal because Reid and I funded Whenere, his start-up effort to build out his own metaverse. What's most impressive is not how right he is, but how committed he is to creating the future he sees coming. In tandem, he's been building Laminaɪ, what he describes as "a digital content protocol for the creator economy of the future."

Finding Your Comic-Con

What's most important is how Neal has gone native, explored edges, and found traction. Neal's approach to getting native on a new platform is to find people already living a primitive version of the future. He goes where the first consumers are. For Whenere, the world-building project that Reid and I backed, he identified Comic-Con attendees and cosplayers—people already blurring virtual and physical reality through role-play, many of whom are his superfans—as test audiences as well as funders, supporters, and creators. That was how Neal deconstructed and found Proven.

"Go to Comic-Con and you see people in costumes from *Game of Thrones, Harry Potter*—whatever. They've spent thousands of dollars; they use vacation time to be with others who love the same world. They're maintaining wikis, tracking every detail. The Hollywood production pipeline—TV, movies, games—can't produce content fast enough for them."

The insight? Don't build for hypothetical future users. Find the weird communities already attempting primitive versions of your vision.

The pioneers who win on new platforms understand it's all about experimentation. "Rather than trying to predict what the metaverse will be," Neal explains, "try to foster an environment where people can try a bunch of stuff. Empower them to try new things. Give them a way to be compensated if their ideas turn out to be profitable."

That's why Neal is working on two levels simultaneously. With Laminaɪ, he's building blockchain infrastructure for creators to get paid for their work in virtual worlds. With Whenere, he's building a specific application: immersive story-world experiences for superfans.

Neal has no use for massive capital either (which is rare to hear). "Sometimes a billion can accomplish less than a million. There are so many cases where people get that kind of capital and end up with layers of hierarchy between them and what's actually happening. Everyone in that structure justifies their salary, and suddenly you're no longer close to the metal."

Neal's biggest insight for all of us pioneering these new planets is to trust your irrational convictions about the future. You don't need to articulate why your vision is right. You just need to believe it enough to build for it. As he puts it, "What do you genuinely believe that others don't?"

So What Do We Do This Week?

One of the hardest things about landing on new planets is finding frameworks to operate within. The search can feel chaotic.

After each of these three interviews, I came away inspired to do something. All three of these people are living life at the speed of play. They're having fun, jamming on projects they're passionate about—and making real progress quickly. My takeaways from the interviews were to have fun and increase our surface area for learning and inspiration.

We see this pattern in all three planet pioneer stories. And I've seen the need for it in my own process as a founder. When I was trying to get Tribe.net right, Sean Parker walked into my office with Zuck and showed me Facebook. They had landed on Planet Social, and they were delivering the winning idea that matched my winning instincts. They created the people web, which became our primary way to have social connections— eventually, even to a fault. Before, we all needed to go out to see people. Today, bars in San Francisco are closing at record rates. The Cocktail Party has gone completely online. I was lucky that Sean showed me this in 2004. But I also foolishly ignored their clearly Proven idea and stubbornly stuck to my own path with Tribe.net.

So, what do we do this week? Leave our houses, our comfortable places. Reid does so by connecting through his amazing network, while getting leverage with people such as Parth Patil, a hacker who is constantly experimenting on his ideas. Cyan seeks out the weird and random every week. Neal goes deep in fantasy communities such as Comic-Con.

Here's my answer for what I'm doing this week. I'm currently seeking new ways to get out of my office and engage. I'm actively working on multiple start-up projects. I'm about to write in my Book of Life again this fall. I'm still hoping this is the year I finally launch Dot Earth; I'm closer than I've ever been.

What matters most is that we engage with our curiosities. What are you drawn to? What do you know the most about? How can you amplify that and bring it to the world in a more amazing way by leveraging new technology?

That is the path to landing on new planets and getting out of the Abyss. It's choppy but not aimless. We can remind ourselves that if we're in the right body of water, we don't have to pick the right boat.

TL;DR

Building for new platforms is hard. Nothing is Proven yet. To build something successful, you have to get native to the new platform. During any platform shift, there are two waves of successful products: the obvious first wave that everyone predicts and the second wave of magical new services you can't imagine until you live there. Give yourself permission to be too early, and hack on experiences at your curious edges—the intersections between your personal interests and emerging technology—until you find something with heat. Increase your surface area to expose yourself to pioneers and accelerate your learnings. Get uncomfortable and start experimenting this week.

How Ambitious Are You?

The first question I have for every founder is this: "How ambitious are you?" Let's be honest. Are you ready to kill your B+ ideas? To operate with intellectual honesty?

Just like the best investors, the best product founders aren't making bets; they're collecting winnings. Any of us would have made the investments in the seed rounds for Google or Facebook. When Sean Parker offered Reid and me the opportunity to invest in Facebook, I thought it was like receiving a winning lotto ticket. (I was way off about how good a winning ticket it actually was!) Similarly, the best product makers know before launch that they have a winning product.

It's not a coincidence that the same founders succeed over and over again. It's not that the world favors them. They're doing something different from everybody else. They changed their odds of success because they got to conviction on both their instinct and their idea. And once they knew they were right, nothing could stop them.

There's something magical about being right—about finding true signal. When you're looking at a restaurant menu and can't decide what to pick, it's because nothing on the menu is right. But when you see the right thing, you just know. It's like that with products. When your product is right, every signal comes back positive. It speaks to *you*. Your friends and family love it. Every anecdote and metric comes back with energy. The product builds its own momentum and starts to take off. Everything just works. And the reality is, when you have the right product, you can do a lot of things wrong and still succeed.

When you chase a B+ idea and you don't have true signal, you get a lot of noise. The problem is that too many of us tune to what we want to see and misread the feedback as reason to stay on this path at the expense of pursuing winning products. This holds us back from failing fast.

Are you playing offense? If you're truly ambitious, you'll act with intellectual honesty to get to true signal, no matter what. And that's what this entire book is about. Every story, every framework, every practice—isolating your winning instincts from your losing ideas, Proven Better New, roadmapping, Bold Beats—all are in the service of getting you to that conviction faster.

It takes incredible ambition and persistence to build products people love. And when you finally produce something that other people connect with, it's one of the greatest highs. That's what I want for all of us.

Your Instincts Are How You Win

After decades of building products, I've learned that your instinct is probably right, but your idea is probably wrong. And that gap between the two is where most founders lose.

The winning path goes through many shots on goal and many failures. Too many founders find an odd sense of honor in sticking with their losing product.

My own example of going down with the ship was Tribe.net. If I could go back in time to 2004, I'd shake myself and say, "Dude, you're on the brink of three new industries, but you're trapped in one losing idea."

The key is to keep developing your instincts—those recurring themes and hunches that continue to haunt you. They're passions, curiosities, interests, beliefs about things you think are missing from or could be added to the world. But you may not know what to do with them yet, because they might not fit in the world today.

You need patience and persistence with these core instincts. But you also need twitchiness, playfulness, and urgency—a bias to action, to try lots of shots on goal without being attached to any single one. The more you can reduce the cost of that shot on goal, the less you have at risk and the more you can experiment playfully. *Sure, I'll try this approach. Sure, I'll build a quick prototype just to see what people think.* I go through this process constantly—trying to figure out how to get my ability to execute cheap enough at the margin so I can keep trying different approaches, learning, and iterating.

We have to get comfortable with a mindset that whatever we're building is probably wrong, but we're going to learn from it. If we take that approach, we don't waste time and cycles building it right and durable—we build something people love, faster.

Start Anywhere

One of the hardest questions is where to start. The reality is, you can start anywhere.

Many hit products started very small with a narrow problem. In 2007, I made a poker game that I thought my friends and I might play. I wondered whether it would catch on Facebook, where everyone was hanging out with nothing to do. That took off so quickly that within weeks, I realized the game could unlock gaming and play for the mass market.

Tobi Lütke was trying to build a website for his surf shop and saw how bad the existing options were. So he built Shopify—which as of February 2026 is a $156 billion company. Melanie Perkins and Cliff Obrecht, the couple who founded Canva, were designers frustrated that no products made beautiful design accessible. So they started Canva—now a $42 billion company.

I have a friend who is the wildly successful founder of more than one of the best-known consumer internet services. Amazingly, he has decided to spend this year treating music like a job, not a hobby. He put together a nine-person band, and they've started recording original songs. They don't even have a name yet. But I ran into a friend who said his set at Burning Man was one of the best he'd ever heard. He committed: He pursued quality for the sake of quality.

These product makers built something for themselves that satisfied their own needs and tastes. The magic happens when we follow the thread of something narrow—a small unmet need or opportunity—that leads to reinventions of whole industries.

An Invitation to Build Treasures

This opportunity is available to all of us. What are you an expert witness about? What are you closer to than anyone else?

Every one of you has the potential to massively change your odds if you start to use these frameworks and accept that the idea you're working on now is probably wrong, but the instinct driving you is probably right.

Mark Pincus's iPhone home screen, November 3, 2025.

Build something you treasure first—something you find value in and are proud of. If you do, the odds are high that other people will, too.

So little of our future has been invented today. When I think of the services that make up our digital life stack—email, phone, weather, calendar, camera, food, travel—I realize it's all waiting for product makers to innovate.

Look at my phone. Half the screen is empty. The half that's filled is mostly occupied by generic apps, waiting for someone to reinvent them. And I'm probably thinking too small. In five years, we might not even use phones. There might be a whole new interface for our digital life stack.

Since the beginning of the internet, there's never been this level of opportunity. It's as wide open as it was in 1994 when

I started my product-making journey, but now there are billions of users generating trillions of dollars. Yes, Internet Treasures—the massive platforms that reshape entire industries—are still being built. Google and search spawned thousands of new businesses. GPT is already doing the same.

By the time you're reading this book, I can't imagine how easy it will be for us to build our ideas. With AI, we're on a path where we can vibe code, vibe create, vibe test, and vibe market products. We're on the brink of living our lives at the speed of play.

No matter what stage of career you're at, consider burning your résumé. Take risks. Spend time following your passions and curiosities and immersing yourself in places where you're really going to learn. Get hands-on experience taking shots on goal around the best people building products in your area of interest. Learn from world-class operators (if you can) before you go off and do it yourself. While you're doing that, you might find your engineers, your cofounders, your people.

Bing told me the best measure of an all-hands meeting was seeing our teams immediately go back to building more amazing products.

My hope is that this book inspires you to go pursue your own ideas right now and to start working on and refining your own taste—what makes products great for you.

These frameworks and practices represent a mindset you can use to approach any idea. By going narrow and small, we find the edges—and that's where we can innovate and expand the world's boundaries.

So, what do you want to build today?

Acknowledgments

I usually skip acknowledgments sections in books, but if you're still reading or listening this far in, maybe you're just in for the whole enchilada.

First, I want to thank my writing partners. Carlye Adler has stuck with this book for over five years now and is one of the only people able to channel me and be an authority on what is my voice. Katie Goldstein, my chief of staff, has been both a drill sergeant and a conduit, channeling many potential readers. Hollis Heimbouch is the best (and only) editor I've ever worked with. She somehow managed to always stay positive while living through a few huge pivots in the focus of our book. I'm appreciative of the entire team at HarperCollins. Jim Levine, our agent, showed so much confidence that this book would find a publisher and an audience that even I started to believe him. Gary Leff, who has been a BFF, has probably read the book more times than anyone else and has been an amazing sounding board and cheerleader. Sarah Selim, for her thoughtful feedback, which was provided speedily before a final deadline. Dan Garon and Kevin Liang, my previous chiefs of staff, kept this project alive and moving in ways that today I can see ensured its survival.

I want to thank Reid Hoffman, who has been my friend and collaborator for decades. His foreword was so good it made me feel like I had to improve the whole book. I also want to thank Reid for modeling what it means to be a true partner and friend.

Bing Gordon, whom I reference many times in this book, has also been a friend and collaborator for almost 20 years. I always say that Bing will give you 10 ideas, and your job is to figure out the 1 or 2 that are such gold you can't afford to ignore them. Bing gave me my MBA and PhD in game design in just two years, and then taught me how to be a founder at scale. (Even though I say fuck scale, it's still a necessary evil.)

Fred Wilson, who enabled me to pursue this crazy journey by taking a crazy flier on my first company, loaning me $250K and spending so many nights on the phone with me in my first seven months that it's also worth thanking his amazing wife, Joanne Wilson, who had to share him with me. Fred taught me that it's all about showing real software and that the shittiest working code beats the slickest fake demo.

John Doerr for naming my why—Internet Treasures—and for being an outstanding human and world-class hugger.

Gina Pell for listening to me go through the entire book in the middle of a dinner party and then telling me that she was repeating my points with her team. Gina became my core audience and use case for the book. She is a brilliant woman who went from being a stay-at-home mom to creating her own start-up.

The amazing entrepreneurs and teachers who both inspired and contributed to this book with interviews and hours of debate: Emmett Shear, Nikita Bier, Cyan Banister, Neal Stephenson, Brian Chesky, Joe Gebbia, Brian Grazer, Shayne Coplan, Josh Davis, Ethan Mollick, and Sarah Torti.

Sunil Paul for giving me my start by cofounding FreeLoader.

Cadir Lee and Scott Dale for teaching me what the best engineers could produce and constantly challenging me across three companies.

My original Zynga family: Eric Schiermeyer, Michael Luxton, Justin Waldron, Colleen McCreary, Marcus Segal, Reggie Davis, Scotty Koenigsberg, Mike Verdu, Henry Stern, David Ko, Brad Feld, Ellen Siminoff, Jon Tien, Mark Skaggs, Eric McDougall (who was and still is the minister of cool), Andrew Trader, Amitt Mahajan, Tim LeTourneau (rest in peace), Steve Parkis, Maureen Fan, Manuel Bronstein, Nick Tornow, Jeffrey Katzenberg, Owen Van Natta, Katie Geminder, Todd Arnold, Barry Cottle, Ian Cinnamon, and Michael Chow.

My Zynga part II family: Frank Gibeau, Bernard Kim, Jeff Shouger, Josh Lu, David Cohen, David Leem, Vee Saghal, Regina Dugan, Lou Lavigne, Carol Mills, Monty Kerr, and Joe Kaminkow.

Matt Ocko for believing in me early, for explaining to the VCs why Support.com's tech was important (and what it could do), and for convincing me in 2006 that I should be going for something much bigger (and eventually becoming my consiglieri once I did).

Craig Newmark for modeling depth of service to customers and befriending Zinga the dog.

Eric Ries for treating me as a writing peer, even though I didn't deserve it, and not minding that I trashed his MVP concept.

I've been one of the most coached people. Thanks to Erika Leder, who has been my life coach for 23 years. Bill Campbell (rest in peace), the legendary Silicon Valley coach, who saw my fire, connected with my heart, and decided to coach me when he had no time or cycles. I already mentioned Bing. And Jorge Bort, who never made me a great tennis player, but did give me wisdom that I could apply to management.

Val Syme and Paul Martino for cofounding Tribe.net. Sorry we didn't succeed despite being early and right.

Nitin Khanna for being an unflappable CTO who can build anything I can imagine, sticking with my crazy Dot Earth vision these past four years, being willing to adapt to my constant pivots, and enduring many moments of deep frustration.

To my beautiful family—Hilary, Georgia, Carmen, Wyatt, Enzo, and Lucia. And my co-parent, Ali Pincus. My mom, Donna Meyers. My dad, Ted Pincus (rest in peace). My four sisters—Laura, Susie, Anne, and Jennifer. My nieces and nephews for being avid gamers and product testers.

And finally, to Tom Cole (rest in peace) for being a brother, teacher, muse, and very first backer (while living on my couch).

Appendix I

Blog Post: Revolution of the Ants

Why i blog—revolution of the ants

Posted on December 10, 2005 (copied here as it originally appeared)

there's been a lot of passionate debate in the comments on this blog about the political topics i've raised like torture and the upcoming execution of tookie williams. i realize that politics has always been considered outside the business realm, and that many of my internet business friends feel it's not their place to comment publicly about their personal politics.

i want to encourage you all to reconsider. we're in the middle of a silent revolution. we've seen it happening in financial markets through shareholder activism which started with lbo's and has progressed to the point where no F500 board room is 'safe' anymore. i applaud carl icahn and lazard for going after time warner and look forward to the day when shareholders are no longer held hostage by corrupt boards.

in fact we've mostly seen the revolution happening in business. it has yet to succeed in the political realm. i've blogged about how my friend matt gonzalez's campaign for sf mayor showed the power of the ants (he lost by 2% with no funding); and about how howard dean's campaign showed the power on a national scale (more donors than any campaign in the history of US politics).

here's my argument why you (internet business friends) should start caring more about the political revolution. the universal public network has enabled massive and immediate swarming. these swarms (like myspace, the facebook, craigslist) can happen overnight, and are reshaping markets.

there is a political undertone to a lot of this. you cant empower people and avoid the question. *you cant let them self govern their community online and avoid the question of offline.* we already see a connection on Craigslist between community values like 'fairness'—as in what is a fair profit for craig— and trust–being wether we should let this site house our activities.

you cant fake it. i dont know what this means for news corp owning myspace. maybe the exception will prove a new rule. maybe i'm seeing the world i hope exists, but i dont think so.

it's only a matter of time before a site comes along that enables the ant swarm to have political power. I guess this sounds like howard rheingold's smartmobs. I don't know the form it will take. But to ignore the political side of sites like craigslist is to only see half the picture.

and that is why i 'blather'on about politics. i will follow the ant revolution wherever i see it with the expectation that one day it will be so pervasive that everyone will join in this imporatant dialog.

one last point. this opportunity goes far beyond money. our great country was freed from one form of bondage in 1776. what we all realize now is that this revolutiononly applied to one class, white and mostly rich. this next revolution will apply to everyone else.

do i have a 'right' to talk about this as one of the white priviledged class? why the fuck not? is it hypocritical of me to 'whine' about this? i'll ask you, isnt it hypocritical if you're not asking this? while i try to be receptive and respectful of both sides, anyone who tries to argue that our country offers equal opportunity to all comers is smoking crack. 'life just isnt fair'. yeah, it wont be fair when you get shot on the street by one of those unfortunates either.

lets start getting our hands a little dirty. lets get out of our comfort zones. dare to be 'inappropriate'. fuck your 'readers'. tell them they get what they pay for. dare to be the first 'yahoo' to cross the line. take a stance. didnt seem to hurt craig.

we have so much power it's scary. more frightening is how seldom we use it for anything that matters beyond our silly web 2.0 world.

"baby, you live in cage with golden bars"

charles bukowski, barfly

Case 2: Use GPT to Build Your Own Hit

Can We Beat Threads?

This is the Proven Better New analysis we asked Chat GPT to do, which we discussed in Chapter 4.

= Proven: Everything Threads Has That Already Works

Core UX Mechanics

- Vertical infinite scroll feed (Instagram + Twitter).
- Standard actions: like, reply, repost, share.
- Threaded conversations (Twitter-proven).
- Inline media (images, videos) with familiar layout.
- Light/dark mode.
- Profile page with avatar, bio, follower/following counts.
- Notifications for likes, follows, replies, and reposts.
- Hashtag-less content discovery (algorithm feed).

Growth and Virality Mechanics

- Instagram identity piggyback (instant onboarding).
- Ability to move followers from Instagram → Threads.
- Cross-app nudges and notifications.
- Seamless account creation (Instagram login).
- Easy posting from mobile.

Safety and Community Mechanics

- IG-scale moderation.
- Brand-safe environment (critical for Meta).
- Clean UX that feels friendly and familiar.

Social Norms

- Short text updates.
- Conversational posting.
- Creator-first discussions.
- Low-pressure posting compared to Instagram.

Copying these is nonnegotiable. Threads already validated all of this at massive scale. This is our starting point.

> Better: 10 out of 10 Users Would Say "Yes—This Is Better"

The goal here is to find areas users already hate on Threads and X where an improvement is obvious and universal—not controversial, not New, simply Better.

The following are high-confidence opportunities.

1. Better feed control (everyone wants this):
   - Give users a true chronological feed + an "Only People I Follow" feed that never resets.
   - Give users a post-hiding and keyword-muting system that's actually effective.
   - Transparent.
   - Expanding that's cleaner than Twitter.
   - "High-signal mode" (auto-filters low-effort replies, spam, rage bait).
   - Quoted replies.
   - Superfast inline editing (no paywall).
   - Auto-captioning for video.
   - Built-in algorithm controls: sliders for "friends," "funny," "news," etc.

   Why it's Better: Every user complains about algorithmic chaos and lack of control.

2. Better conversation tools:
   - Thread collapsing/expanding that's cleaner than Twitter.
   - "High-signal mode" (auto-filters low-effort replies, spam, rage bait).
   - Quoted replies shown visually—easier reading.

Why it's Better: Reply stacks are a pain on Threads and X. Zero disagreement: All users want a cleaner conversation UI.

3. Creator utility:
   ○ Superfast inline editing (no paywall).
   ○ Auto-captioning for video.
   ○ Built-in link unfurling that's prettier than OG cards.

Why it's Better: Every creator wants friction removed.

4. Search that actually works:
   ○ Real-time search.
   ○ Trending topics with context cards (but without the toxicity).

Why it's Better: Threads search is bare-bones; X search is messy. A middle ground is obvious.

+ New: Novel Ideas Never Tried with This Audience

This is the riskiest part—so each "shot on goal" below isolates the New so it cannot confuse signals.

Shot on Goal #1

+ New: The "Room" Model—Private Public Spaces

A new social mechanic: rooms that are public to the world but *separate* from your main identity:

- Posts inside a Room don't appear on your main profile.
- You can "enter" or "exit" Rooms with one tap.
- Rooms have shared vibe settings (funny, analytical, visual, etc.).
- Creators can run Rooms effortlessly—zero setup friction.
- Rooms give structure without the heaviness of Reddit communities or Discord servers.

This is Reddit × Twitter × Discord, simplified for everyday mobile users.

Why This Shot Has a Good Chance of Succeeding

1. It's built on top of a proven core (feeds, posting, profiles). We're not changing the "slot machine" mechanics—we're adding a new layer on top of the proven Threads/Twitter interaction model. This dramatically lowers risk.

2. It remedies an existing, well-known pain. Users want topic-based spaces without sacrificing their main identity or drowning in algorithmic sludge. Rooms isolate conversations in a way that X and Threads simply don't support.

3. It involves zero friction for users and creators. Reddit and Discord require setup and admin overhead; Rooms feel like stepping into a casual "conversation zone." Low friction = rapid adoption.

4. It is highly screenshot-able. Room launches, inside jokes, Room leaderboards—all produce screenshot loops that drive growth organically.

5. It fits mature-market dynamics. Social is a crowded space; you need a bold-but-simple New mechanic. Rooms are exactly that—a new social structure users intuitively understand.

Shot on Goal #2

+ New: Social "Draft Mode"—Post Together, Privately First

A brand-new social mechanic that brings friends *into* your posting workflow:

- Start a post → select one to three friends → they see it in Draft Mode.
- They can suggest edits, add images, react, or riff.
- You publish the final version publicly in one tap.
- Optional: show "cocreated with X" badges.

This turns posting into a shared creative moment—social creation, not solo broadcasting.

Why This Shot Has a Good Chance of Succeeding

1. It builds on Proven posting mechanics. Draft Mode doesn't change the feed, UI, or posting flow; it adds a lightweight step on top of what already works. Risk stays isolated.

2. Posting anxiety is universal. Users often hesitate to post because they don't know whether something is good. Having a friend gut-check—privately—unlocks *more frequent posting*, exactly what early-stage networks need.

3. It creates a new category of social behavior. Cocreation is a massive emotional unlock. Users feel closer to their friends through shared creative moments.

4. The output is screenshot-worthy and viral. Behind-the-scenes drafts, friend reactions, cocreated posts: all prime screenshot material.

5. It increases stickiness for creators and users. Creators get an easy editing layer. Regular users get confidence and social belonging. Both groups benefit—giving the mechanic broad appeal.

Shot on Goal #3

+ New: "Live Threads"—Real-Time Group Posting

A synchronous posting format:

- A Live Thread opens for a two-minute or five-minute window.
- Everyone posts simultaneously.
- The final product becomes a collage of text, images, and reactions.
- Live Threads can be public (events, memes) or private (friends, campus groups).

Think BeReal energy + TikTok Live dynamics + Twitter threads—but reimagined as a feed-native mechanic.

Why This Shot Has a Good Chance of Succeeding

1. It's layered on top of Proven mechanics. Feed remains the same. Posts remain the same. Live Threads are a temporary "mode," so the New is cleanly isolated.

2. It creates real-time magic—something existing platforms lack. Twitter/X is real-time but not synchronized. Live Threads make posting feel like an *event*, not a broadcast.

3. High emotional payoff. Synchronized posting produces hype, anticipation, and group identity—a powerful combo missing from current social products.

4. Hugely screenshot-able. A Live Thread from a party, sports moment, meme storm, or campus event is instantly shareable.

5. Perfect fit for a mature market. When markets mature, you need bold, simple New mechanics—OMFG moments. Live Threads are exactly that: familiar enough to understand instantly, new enough to feel magical.

Glossary of Terms

Abyss

This is the space when you are not actively working on something or when products have not yet achieved product market fit. My dad used to call these states being "in between successes." The Abyss is a painful period of time marked by observing, trying things, and failing. Don't rush past these moments. They're when you develop the conviction that will sustain you through the years of iteration required to find the winning idea.

Active Social Network (ASN)

Users who are regularly engaged and connected to others within a product or platform. ASN metrics help measure the health of social features and viral growth mechanisms

atomic unit

This is the smallest meaningful unit of a product or idea that can be tested. In roadmapping, the atomic unit of the Blue Sheets was usually a new feature. The concept emphasizes testing and failing fast with the minimum necessary to validate an idea, rather than building a full product before getting feedback.

B+ is the enemy of an A

This principle is a warning against settling for "good enough" when pursuing greatness. B+ ideas feel safe and achievable but prevent teams from reaching for truly exceptional A-level work. This principle encourages taking the risks necessary for breakthrough innovation rather than incremental improvement.

Blue Sheets

The roadmap meeting became the core engine of everything at Zynga. I made a top-down call that changed everything: "Everyone is going to work off the same roadmap document." It became known as the Blue Sheets. That was what turned the Blue Sheets into a competitive advantage; it enabled us to adapt and optimize improvements between and across teams. The atomic unit of the Blue Sheets was usually a new feature.

boil the ocean

This is what you do when you attempt overly ambitious projects that try to solve too many problems at once—the opposite of focused, achievable Bold Beats. Teams should avoid trying to boil the ocean and instead focus on targeted innovations that can actually ship and make an impact.

Bold Beats

A Bold Beat is a positive disruption in the consumer experience. Bing Gordon called them OMFG moments: They light up social media channels. At Zynga, we said that something was a Bold Beat when 25 percent of users clicked on it and it moved key metrics by 10 percent or more. Bold Beats sit between peanut-size tweaks (too incremental) and boiling the ocean (too ambitious). The most powerful Bold Beats evolve into Golden Mechanics.

Broken Résumés

These are career paths that don't follow conventional trajectories. A Broken Résumé often indicates someone who has taken risks, tried different things, or pursued their passions rather than optimizing for résumé polish. It can be a positive signal of entrepreneurial thinking and adaptability.

Cocktail Party

We are all searching, online and off-line, for that perfect Cocktail Party—the gathering that combines curated people with the right context, purpose with fun, and trust with community. It's where interesting people, old friends and new, come together effortlessly. It's where we find our best "leads" for everything from dates to jobs, travel ideas to culture. Reid Hoffman calls this "the people web." This instinct is at the core of some of the biggest internet companies and industries, from Meta to TikTok.

code red

This is a critical situation in which a team or product is significantly underperforming. In roadmap meetings, I could identify a code red situation by seeing missed expected outcomes and inaccurate engineering day estimates. This situation indicates a need for intervention and course correction.

Daily Active Users (DAU)

"DAU" refers to the number of unique users who engage with your product each day—a key metric for measuring product engagement and growth. This metric is often paired with MAU (Monthly Active Users) to understand usage patterns.

Day 365 (D365) retention

This is the percentage of users still active one year after first using your product—a powerful long-term retention metric that indicates product stickiness and lasting value. It shows whether users form lasting habits with your product.

deconstructing

This is the process of breaking down an existing product or feature to understand what makes it work. It is part of the Proven Better New framework—deconstructing successful mechanics from other products to identify what to bring into your own innovation. This process helps separate the Proven elements from the New innovation you want to add.

engineering days (eng days)

A unit of measurement for engineering effort required to build a feature. Teams estimate engineering days needed for each feature in the roadmap and track actual versus estimated eng days to improve accuracy over time. This metric is essential for understanding the cost and benefit of each feature and maximizing return on engineering investment.

expected outcomes (EOs)

At Zynga, each game studio team had to show up with a hypothesis they were testing that was tied to their objectives, expected outcomes per engineering day (straight from their Blue Sheets), real metrics versus those expectations, and what they learned and how it would shape next week's bets. Teams commit to expected outcomes (e.g., an additional 150,000 installs per day, an additional $50K in revenue per day) in roadmap meetings, then compare actual results afterward.

expert witness

This is a frustrating role in corporate jobs: You're closest to the answer but furthest from the decision. The role represents the dysfunction of organizations where decision-making power is disconnected from domain knowledge. As I like to say, "corporate" means lack of thought or intent!

FarmVille

Zynga's breakout social game became one of the most successful games of all time, reaching more than 80 million MAU. FarmVille demonstrated the power of social gaming mechanics and exemplified many of the principles in this book, including Bold Beats (e.g., "animals that move"), Golden Mechanics, and continuous innovation through roadmapping.

focus group of one

This refers to using one person, which can be yourself, as the primary user for testing and validating product ideas.

forever franchises

A strategic, long-lifecycle product (often, a gaming title) designed to entertain users for many years through continuous updates, live ops, and evolving content. These franchises are designed to generate high, consistent revenue—often $100M+ annually—and create deep engagement characterized by a 10 percent or higher D365 retention.

full-stack leader

This is a product leader who can operate effectively across all aspects of product development—from user insight to technical implementation to business metrics. Such a leader can move between high-altitude strategy and ground-level execution, and has connected all the dots through hands-on experience.

ghost ship

This is a product or feature without clear ownership or direction, drifting without purpose. This kind of problem often happens when a founder leaves the company. Founderless is rudderless.

Golden Mechanics

These are proven social game mechanics, which became core building blocks of every subsequent game at Zynga. When you find a Bold Beat that truly resonates, you can turn it into a repeatable pattern that drives sustainable innovation. It becomes part of your innovation path, a proven mechanic you can build on again and again.

heat

This is player excitement and engagement that drives usage and retention. The goal of product development is to find heat and transform it into things users love, which translates into business metrics.

imagination chessboard

This entails envisioning what you want life to look like two or five years out—and then considering all possible moves to help you get there.

Internet Treasures

These are services we can't remember life before or imagine life without. John Doerr and Bing Gordon believe that one day, these products will be displayed in the Smithsonian as the greatest output of our culture: the iPhone, Amazon, Google, Instagram, Facebook, TikTok.

instinct veins

Instinct veins are deep sources of insight about human needs and behavior that can spawn multiple product ideas, even whole new industries. These veins are so fundamental that you can return to them again and again, each time extracting new possibilities

Minimum Idea State (MIS)

This is the smallest expression of an idea needed to test its core value proposition. It goes beyond Minimum Viable Product to focus on validating the idea itself before building, using prototypes, ripomatics, and fast tests to reach conviction about an idea with minimal investment.

Monthly Active Users (MAU)

This is the number of unique users who engage with your product in a month. This metric is paired with DAU to understand engagement patterns and calculate DAU/MAU ratio, which indicates how often users return to your product.

moral contract

This is an unwavering bond with every member of your team that, if they deliver, you will recognize and reward their contribution, often with a bigger role and more compensation. But the moral contract goes beyond this transactional definition: You've got their backs. The bond is based on the premise that we are all going to war together.

mouse nuts

These are tiny, insignificant changes that don't move the needle. Teams in execution mode often default to mouse nuts when they should be taking bigger swings.

objectives and key results (OKRs)

Objectives define what hill you're taking (important, knowable, achievable), and key results are measurable outcomes that prove you've succeeded. We used this system to define the level of growth, risk, and innovation we expected every team to sign up for every quarter. I gave a three-hour lecture on OKRs at Harvard Business School, which I titled "OKRs: A Love Letter from the CEO." OKRs invite innovation and autonomy, while also driving accountability and growth. They also give the CEO a framework to manage and course correct.

OMFG moments

These are features that create surprise and delight and light up social media. Examples include animals that move in FarmVille (built in half a day by an engineer) and avatar babies in YoVille. These moments generate player excitement and often unlock new dimensions of gameplay that teams hadn't considered.

Penguin Effect

Cadir Lee coined this term to explain how hard it was to get teams "to walk off their cliffs" and innovate. But the minute another team tried something new and it worked, you didn't have to tell anyone else to try it; it went Zynga-wide within an hour.

Proven Better New

This framework is a systematic approach to generate winning products faster. By protecting us from our urge to make everything new, we avoid wasted cycles and generating false signals. Proven Better New also helps us avoid the time to reinvent necessary features. In other words, we need to get comfortable copying what already works (legally), so we can spend all our time and energy on the novel innovation that actually excites us and our users.

ripomatics

These are prototypes created by combining screenshots and recordings from existing products rather than building from scratch. With AI and vibe coding, teams can now create interactive ripomatics that approach fully functioning prototypes.

roadmapping

Your roadmap is where strategy meets execution. It's the tool that determines which hills we take, what outcomes we expect, and how much engineering effort the strategy is going to cost. It's a living weekly report card. At Zynga, roadmapping was how we ran our factory. It allowed us to translate raw heat from ideas into features that had clear, measurable player outcomes.

Rule of Four

Everything costs twice as much and takes twice as long as expected. If you apply this rule to building mouse nuts features—small, incremental changes—you've just wasted the whole quarter on something that won't move the needle.

shots on goal

These are attempts at creating successful features or products. The key is not just taking more shots, but taking better shots—which requires having more ideas in the funnel to choose from.

sinking speedboat

This is a fast-moving product or feature that's fundamentally flawed and headed toward failure. Speed and vitality don't matter if you're going in the wrong direction or the boat is taking on water and not keeping people retained or engaged.

social breadcrumbs

These are a way to bring players back to the game.

squeezing the lemon

This is the misguided practice of prioritizing monetization and annoying pop-ups rather than addressing player complaints or bold new features.

staying close to the metal

Any good CEO maintains a direct connection with their product and users—what I call staying close to the metal. In most cases, we're talking about software, but this could also be whatever the atoms are that make your product. Staying close to the metal is about channeling our customers and forging our products from atoms and pixels into magical experiences.

stop time

This means pausing to reflect deeply on a problem or opportunity and consider what our future self will thank us for. Committing to a vision with specific outcomes is a way to stop time. I always say to my teams, "We need to stop time, because this _____ is so important."

talent myth

Everyone tells you that your first 10 or 20 employees have to be amazing. It's not incorrect that people matter, but the advice creates an unrealistically high bar. You need people who are available and actually want to join you. Look for people who made failed versions of products with technology that you like; you get these weird, quirky people who can become amazing team members and whom you'd never normally meet.

taste

Taste is the ability to recognize great product experiences and make decisions that lead to quality. This isn't about having universally appealing taste; it's about developing a keen sense of quality and maintaining high standards. A product maker who doesn't use and deeply understand their own product is like a chef who doesn't eat their own cooking.

teaching hospital

At Zynga, our teaching hospital helped us build an organization that could learn and adapt faster than anyone else. The combination of our ruthless metrics tracking through Blue Sheets and our teaching culture turned us into a living laboratory, where every feature launch was a chance to get smarter as a company.

time machine

A Book of Life is a time machine that can go backward and forward in life. This practice builds a hunger to disrupt your own patterns. If you come back in time from five years in the future, what will you thank yourself for doing?

true signal

True signal is about tuning into that exact frequency that speaks to all of us. You feel it inside when a product nails it.

Whales

These are high-value users who spend significantly more money than average users. In free-to-play games, a small percentage of whales often generate the majority of revenue. Understanding and serving whales while not alienating other players is a key balance in game design and monetization.

wildcat drilling

This is taking significant risks to find new opportunities, similar to drilling for oil in unproven territory—not doing anything that's Proven and therefore taking on the most risk, as all New (usually) fails.

About the Author

Mark Pincus is a legendary tech entrepreneur who has spent his career chasing Internet Treasures—products people can't imagine living without. He is the founder of Zynga, a pioneer of social and mobile gaming that created the playbook for viral consumer products. Prior to Zynga, he founded Tribe.net, a social network that preceded Facebook; Support.com, which he took public; and FreeLoader, which was acquired after seven months. He was a founding investor in category-defining platforms such as Facebook and Twitter, and later cofounded the investment firm Reinvent Capital, which took several companies public at multi-billion-dollar valuations. He also created the product management class at Stanford Business School. Mark graduated summa cum laude from the Wharton School and earned his MBA from Harvard Business School. His greatest creation is his five amazing kids, who teach him every day about the magic of play.